DATE			

ROBOTIC MANIPULATION STRATEGIES

Michael A. Peshkin

Northwestern University

Prentice Hall

Englewood Cliffs, New Jersey 07632

Library of Congress Cataloging-in-Publication Data

Peshkin, M. A.
    Robotic manipulation strategies / Michael A. Peshkin.
        p. cm.
    "Based on a dissertation completed in November, 1986, for the PhD
degree in the Department of Physics, Carnegie-Mellon University,
Pittsburgh, Pennsylvania"—Acknowledgements.
    Includes bibliographical references.
    ISBN 0-13-781493-3
    1. Robots—Dynamics. 2. Manipulators (Mechanism) I. Title.
TJ211.4.P47   1990
629.8'92—dc20   89-26580
                                                                  CIP

R E F

Editorial/production supervision
  and interior design: *Jane Bonnell/Gertrude Szyferblatt*
Cover design: *Lundgren Graphics, Ltd.*
Manufacturing buyer: *Ray Sintel/Kelly Behr*

The publisher offers discounts on this book when ordered
in bulk quantities. For more information, write:

Special Sales/College Marketing
Prentice-Hall, Inc.
College Technical and Reference Division
Englewood Cliffs, NJ 07632

Printed in the United States of America

10 9 8 7 6 5 4 3 2 1

ISBN   0-13-781493-3

PRENTICE-HALL INTERNATIONAL (UK) LIMITED, *London*
PRENTICE-HALL OF AUSTRALIA PTY. LIMITED, *Sydney*
PRENTICE-HALL CANADA INC., *Toronto*
PRENTICE-HALL HISPANOAMERICANA, S.A., *Mexico*
PRENTICE-HALL OF INDIA PRIVATE LIMITED, *New Delhi*
PRENTICE-HALL OF JAPAN, INC., *Tokyo*
SIMON & SCHUSTER ASIA PTE. LTD., *Singapore*
EDITORA PRENTICE-HALL DO BRASIL, LTDA., *Rio de Janeiro*

Contents

Preface

In this book I treat the problem of predicting (and exploiting) the sliding motions that workpieces undergo when they are manipulated on a surface by a robot, or when they pass through a parts-feeder. The results are useful in the automated design of parts-feeders and of robot manipulation strategies.

The prediction problem, although it falls squarely in the domain of classical mechanics, is nontrivial because it is not enough to have predictive power for a well-described initial condition. We need bounds on the possible motions of a pushed workpiece for a broad class of initial conditions. Specifically, the distribution of pressure between a workpiece and the surface it slides on is realistically unknown, so the motion of the workpiece in response to a push cannot be predicted uniquely. In Chapter 3 the set of possible motions of a workpiece for a given push, for all collections of points of contact, is found. The answer emerges as a locus of centers of rotation.

Even when a prediction problem (such as the sliding problem) is fully understood, automated synthesis (also known as "planning" or "design") may be intractable. The synthesis problem here is to use sliding motions, without sensors or feedback, to orient and align workpieces from an initial random state. *Configuration maps* are introduced, mapping all configurations of a workpiece before an elementary sliding operation to all possible outcomes, thus encapsulating the physics and geometry of that operation. Using products of configuration maps and appropriate search techniques, operation sequences can be found that reduce the configurational uncertainty of a workpiece. As an example, in Chapter 5 we design automatically a class of passive parts-feeders consisting of multiple sequential fences across a conveyor belt.

Along the way I describe a simple variational principle for constrained quasi-static mechanics which I call the *minimum power principle* (MPP): "a constrained quasi-static system performs that allowed motion which minimizes the instantaneous power.'' The principle seems intuitive, but it is false. Only under certain quite restrictive conditions is the minimum power principle true, and these are not uncommon in robotics.* The minimum power principle is considered in Chapter 2.

Michael A. Peshkin

* I am aware of several authors who have recently done or are currently doing further work on MPP and related ideas: Jeff Trinkle (Arizona), Imin Kao and Mark Cutkosky (Stanford), and Suresh Goyal and Andy Ruina (Cornell).

Acknowledgments

This book is based on a dissertation [54] completed in November, 1986, for the Ph.D. degree in the Department of Physics, Carnegie-Mellon University, Pittsburgh, Pennsylvania.*

I wish to thank my thesis advisor, Arthur Sanderson, for the proper combination of guidance and freedom and for getting me quickly into the mainstream of robotics. Matt Mason introduced me to the problem of manipulation of sliding objects. I benefited greatly from discussions with him. I also thank Robert Schumacher for insisting that the theoretical mechanics problem addressed in Chapter 2 be considered, and for several discussions about friction.

My change from experimental physics at Cornell to robotics at Carnegie-Mellon would have been impossible without the flexibility and accommodation of the CMU Department of Physics headed by Raymond Sorenson. I especially appreciate the help of Robert Griffiths in this regard. The Robotics Institute at CMU provided a stimulating atmosphere in which to learn and work.

Michael Fuhrman blazed the trail in the physics/robotics combination. James Russ served as committee chair for both of us. I am grateful to them as well as to Lee Weiss and Nigel Foster for advice.

Gerard Cornuejols, Jeff Koechling, Luiz Scaramelli Homem de Mello, Mark Cutkosky, Matt Mason, Marc Raibert, Robert Schumacher, Yu Wang, and Randy Brost provided constructive criticism of my papers and talks, and Mark Zaremsky and Nigel Foster helped familiarize me with robots and with the local computers. Thanks to Jim Schubert for quick work in the machine shop.

* Several equations were simplified using MACSYMA. This work was supported by a grant from Xerox Corporation and by the Robotics Institute, Carnegie-Mellon University.

    The writing of the thesis upon which this book is based would not have been possible without the cooperation of my daughter Danielle, who slept through the night at a young age. Finally, I wish to thank the Peshkin and Strandburg families, and most of all my wife Kathy, for encouragement, love, and faith in me during my graduate studies.

1

Introduction

1.1. PLANNING WITH UNCERTAINTY

Given geometric models of a robot, workpiece, and environment, and clearly expressed goals, automatic planning of robot manipulation is still a very difficult task. One difficulty is that some discrepancy between model and reality is unavoidable, and a good plan must be sufficiently robust to succeed despite the errors.

One has a choice of philosophies with regard to planning in the presence of uncertainty. One can make plans based on the model, and hope, or preferably check, that these plans succeed in the presence of typical errors. In a well-controlled environment, or for a model sufficiently updated with sensory information, this philosophy may be adequate.

Alternatively, one can include uncertainty in the model and explicitly plan for its control. This approach does not necessarily make planning more difficult. Consider planning the motion of a cylindrical mobile robot in a cluttered room. Projecting the robot and the obstacles onto the plane of the floor, one can shrink the robot to a point and expand the obstacles to compensate, reducing the problem to that of moving a point [48] [66]. Since the shortest path from start to goal would skim the obstacles, which is undesirable, we could pick a path equidistant between obstacles. This plan may succeed and can be checked. Or uncertainty may be incorporated explicitly in planning by further expanding the obstacles by the uncertainty in their position and in the robot's position. Then the shortest path will succeed by construction and need not be checked.

1.1.1. Physics and Uncertainty

One important component of planning with uncertainty is reasoning about the physics of the interaction between robot, workpiece, and environment. The foregoing example is perhaps unrepresentative because it was only necessary to make plans which avoid the positional errors, not plans which control or eliminate them. In grasping, in manipulation, or in assembly operations, our knowledge of the position of a workpiece is changed by an operation. These operations necessarily include contact between robot, workpiece, and environment. Therefore the physics of bodies in contact, rather than purely geometry as in the path planning example, must be considered.

Chapters 3 and 4 address this issue. The physics problem considered is that of a rigid body sliding on a planar surface. While by no means the only nontrivial mechanics problem which must be addressed in manipulation planning, it is nevertheless an important one, first considered in the context of robotics by Mason [44]. When a workpiece rests on a tabletop, it rests in planar contact, and any attempt to grasp or push it may result in sliding. This sliding may increase or decrease our knowledge of the workpiece's location and orientation, as we will see shortly.

Besides positional errors, another form of uncertainty which breaks the connection between model and reality is lack of knowledge of the workpiece. A robot must deal with many instances of a workpiece, and these may vary not only from the model but from each other. The motion of the workpiece may be determined by rather fine details of its manufacture.

Consider a workpiece which has a nominally flat surface. This surface may be in contact with another flat surface, that of the tabletop. The actual points of contact between workpiece and tabletop can depend on particles of dirt, or on slight deviations from flatness of either surface. We will find the set of all possible motions of such a workpiece when it is pushed, without knowing the details of the planar contact. This allows planning when we cannot know the details, as in the case of nominally flat surfaces. It also allows us to plan in ignorance of the details, even when they could in principle be known, as when the workpiece's surface has three screw heads protruding from it.

We can conclude from this example that when there is uncertainty, it is not enough to be able to calculate motion for any completely defined situation. For manipulation planning we need a deeper insight into some kinds of mechanics (such as sliding) which allow us to find sets of motions for a range of instances.

1.1.2. Geometry and Uncertainty

The other component of planning in the presence of uncertainty is geometric (rather than physical) reasoning. Here we have to consider which surfaces of modeled bodies will come into contact, given some uncertainty of the configuration of a workpiece and of the actual motion performed by the robot. Geometric reasoning determines the possible parameters of contact. Physical knowledge is needed to determine the possible outcomes of each possible contact. The geometry of the problem is in turn altered by the outcome of the interaction. Planning requires an integration of geometric and physical reasoning to synthesize a control strategy.

In Chapter 5, we show how planning may be done on the basis of a representation of an elementary operation we call a *configuration map*. The configuration map describes simultaneously the geometric and physical consequence of one operation on all (or a range of) initial states. In our approach the *operation*, with its geometric and physical consequences precomputed, is the basic element of planning.

1.1.3. Sensing

Another approach to dealing with uncertainty would use information gathered from sensors to modify motion plans during execution. Besides the additional hardware and software complexity it introduces, reliance on sensors alone is not a sufficient solution. First, at some level of resolution any sensor fails. Below that level it is the physics of contact between bodies that determines the outcome of an operation. Second, and perhaps more fundamentally, even given complete information about the present state, which is the most sensors can hope to provide, we still have to plan how to get from that state to the goal. In the case of the sliding workpieces which are the subject of this book, we may know exactly where a workpiece is, but still not know how to push it to get it where we want it to go, because we do not know how it will respond to a push. Here too, at some level the intrinsic mechanics of a task takes over from control based on sensing.

We might suppose that in sufficiently sensor-intensive environments little uncertainty will remain in the geometry of contact between objects. If so, the physics of contact can be decoupled from the geometry, and much of the motivation for combining the two in a configuration map will disappear. However, in the case of sliding motion, a "sufficiently sensor-intensive" environment might have to detect the points of contact between surfaces, which is difficult. In the opposite extreme of completely sensorless manipulation, a great deal of uncertainty in geometry exists, and to try to plan manipulation strategies by repeated appeals to geometric and physical simulators becomes impractical.

Although the strategies discussed in this book are completely sensor-less ones, there is no reason why similar strategies cannot be used in combination with sensors. There are at least three areas in which such integration could be considered. First, the mechanics of contact can be used to plan a motion to take a workpiece from a configuration identified by sensors toward a goal configuration, thus allowing less frequent sensor readings. Second, sensorless strategies can take over when the resolution limit of sensors is reached. And, third, sensors can be used to limit the initial uncertainty with which sensorless strategies must contend. This subject is discussed further in Section 7.1.

1.2. EXPLOITING MECHANICS IN ASSEMBLY

Sometimes it is possible to take advantage of contact forces to accomplish a task in the presence of uncertainty. This may be done using compliant robot mechanics.

1.2.1. Compliant Mechanics

A common task which requires physical understanding of object contact is close-tolerance parts mating and, specifically, the peg-in-hole problem. The problem is difficult because the tolerances needed to avoid jamming are often smaller than the positional accuracy of the robot, or the accuracy with which the positions of the peg and hole are known, or both. Humans solve the accuracy problem by using force feedback, modifying their motions in accordance with forces detected. Robots can use this technique too [25] [49] [69], but alternatively they can be fitted with a compliant wrist whose motion in response to forces is appropriate for peg insertion [70] [72] [16]. The point in space (if such exists [34]) at which the matrix relating forces and torques to displacements and rotations becomes diagonal is the center of compliance. Placement of the center of compliance ahead of the end of the peg facilitates peg insertion.

In the technique illustrated in Figure 1-1 (copied from Erdmann [17]), the robot is commanded to move the peg down and to the right, diagonally. The point of first contact is arranged to be well to the left of the hole. After first contact, the peg slides to the right until it drops into the hole. This operation takes advantage of the mechanical interaction of the part held by the robot and the workpiece.

It is required that the robot be able to execute a *compliant move* in the commanded direction. (Mason [41] contains a useful treatment of compliance.) We command the robot to move down and to the right, but when it encounters a resisting force in the $+y$ direction, it should continue its motion to the right, while exerting a predetermined maximum force in the $-y$ direc-

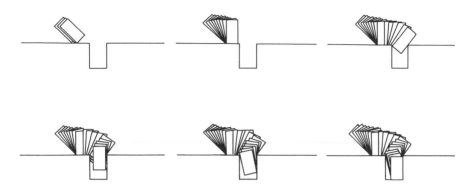

Figure 1-1   Erdmann's peg-in-hole simulation with compliant motion, from [17].
The robot is commanded to move the peg down and to the right, diagonally. The
point of first contact is arranged to be well to the left of the hole. After first
contact the peg slides to the right until it drops into the hole. This operation takes
advantage of the mechanical interaction of the part held by the robot and the
workpiece.

tion. Raibert and Craig [59] described a hybrid force/position control which
is useful for compliant moves.

1.2.2. Sliding Friction

Both of the techniques just described depend on compliance of the
robot. Without compliance, position control combined with uncertainty in
workpiece location can lead to uncontrolled forces when contact is attempted.

Most robot controllers are incapable of compliant move; they are
purely position controlled. This is unfortunate from the point of view of
exploitation of task mechanics, which necessarily involves contact between
workpieces. Since the positions of the workpieces are not known exactly,
many operations (for instance the diagonal compliant move in Figure 1-1)
cannot be performed under pure position control.

The types of operations discussed shortly, and those which are the
subject of this book, can be performed under pure position control, and
without special compliant robot mechanics. A common characteristic of
these techniques is a reliance on sliding. In the examples just given a controlled force was needed to perform operations involving contact in the
presence of uncertainty. A position-controlled robot can produce a controlled force by making use of sliding. Sliding friction translates a controlled
movement into a controlled force, which the robot could otherwise not
produce. That force can then be utilized in manipulation strategies in much
the same way as if it were produced by force control or compliant robot
mechanics.

1.3. EXPLOITING MECHANICS IN GRASPING

Important criteria for planning a grasp are uncertainty reduction, grasp stability, and grasp strength. There is no reason to think that these criteria must be (or even in general can be) met in a single operation.

1.3.1. Uncertainty Reduction

Grasping is often the crucial first step at which uncertainty in a workpiece's position and orientation is encountered. Many robotic operations involve a workpiece which is free to slide on a tabletop or a conveyor belt. Individual strategies have been developed which take advantage of sliding friction between the workpiece and the surface it slides on to facilitate accurate positioning of the workpiece or reliable grasping of it.

An example is the hinge-grasp strategy used by Paul [57] (Figure 1-2). To understand this and similar operations, Mason determined the conditions required for translation, clockwise (CW) rotation, and counterclockwise (CCW) rotation of a pushed workpiece [44]. Mason's result is summarized in Section 1.6.2.

Another example is the strategy of centering and aligning a workpiece to be grasped by first squeezing it with the robot's gripper. This operation has been analyzed in detail by Brost [12]. Brost's results enable the automated planning of a grasping motion (consisting of a translation of the robot hand while the gripper is simultaneously closed at a controlled speed) which will acquire the workpiece in a uniform way despite the presence of some uncertainty in the workpiece's initial orientation.

1.3.2. Grasp Stability and Grasp Selection

Finding a geometrically feasible grasp from a model of a workpiece, even for simple grippers, is a nontrivial task. Several authors consider selection of grasping surfaces from a geometric model [35] [32] [55]. Salisbury [63] discusses grasp selection in terms of constraining the grasped workpiece. Asada and By [3] have done recent work on selection of contacts to allow detachment of a grasped workpiece.

Which of the feasible grasps is then "best" is not well defined, since different tasks, materials, and presentations of the workpiece may require different grasps. Various physical properties of grasps are clearly relevant. Wolter, Volz, and Woo [73], Jameson [26], Jameson and Leifer [27], Barber, Volz, and Desai [5], Holzmann and McCarthy [24], and Li and Sastry [33] calculate the resistance of the grasp to forces applied to the workpiece. Cutkosky [14] [15] considers whether the workpiece will become unstable as gripping force is increased. Asada [2] and Baker, Fortune, and Grosse [4] study the local stability of the orientation of the workpiece in the grippers, in

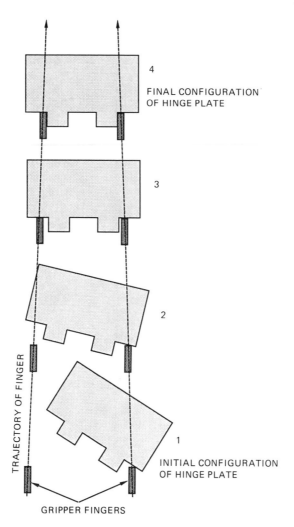

4

FINAL CONFIGURATION
OF HINGE PLATE

3

2

TRAJECTORY OF FINGER

1

INITIAL CONFIGURATION
OF HINGE PLATE

GRIPPER FINGERS

Figure 1-2   Hinge-grasp strategy (Paul [57] and Mason [42]). The robot fingers follow the trajectory indicated by the dotted lines, closing as they translate. On contact with the hinge-plate the trajectory causes the plate to rotate into alignment with the gripper and then to self-center. This open-loop strategy requires no sensing and succeeds despite some uncertainty in the initial configuration of the plate.

the absence of friction. Fearing [19], Cutkosky [15], Kerr [29], and Nguyen [47] consider stability and constraint including friction. Trinkle [65] considers the problem of finding an enveloping grasp for lifting a workpiece away from a surface, including consideration of friction.

Analysis of the maximum torque a grasp can sustain has a difficulty in common with analysis of the motion of a sliding workpiece: the distribution of forces in planar contacts is usually not known, as it may depend on fine details of the contacts. Barber, Volz, and Desai [5] deal with this problem by assuming a linear force distribution. The force distribution affects the torque required to rotate the two planar surfaces relative to one another. The sliding problem is solved without information about the force distribution in Chapter 3. The solution is probably applicable to calculation of the torque resistance of a grasp as well.

1.4. WORKPIECE ALIGNMENT AND PARTS-FEEDERS

Parts-feeders perform uncertainty reduction without grasping. Boothroyd [7] has analyzed and cataloged classes of parts-feeders. Recent work by Lozano-Perez [38] and Natarajan [46] begins to build a systematic understanding of parts-feeders. Automated design of a class of parts-feeders is demonstrated in Chapter 5 of this work.

Mani and Wilson [40] constructed a general-purpose parts-aligner based on sliding friction. A workpiece, initially having unknown orientation, is pushed across a tabletop by a straight fence. The angle of the fence and its direction of motion are under computer control, making the aligner a sort of special-purpose robot. By a properly selected sequence of pushes, the workpiece can be brought into a unique final orientation independent of its initial orientation. The planning method described in Chapter 5 can be applied directly to Mani and Wilson's system.

Erdmann and Mason [18], inspired by work of Grossman and Blasgen [23], devised a system in which a workpiece dropped at random into a rectangular tray is aligned by a sequence of tipping motions of the tray. With each movement, the workpiece slides across the bottom of the tray, or slides along one edge of the tray, or both. Automated planning of the sequence of tipping motions required to orient a given workpiece has been demonstrated [18], and the system has been tested by using a robot to tip the tray.

1.5. STATE SPACES, OPERATION SPACES, AND PLANNING

Much effort has been expended on the automatic planning of manipulation strategies [60]. Several such works have dealt with uncertainty: Lozano-Perez, Mason, and Taylor [37] developed a method using compliance and explicitly dealing with uncertainty. Erdmann [17] developed algorithms for planning manipulation strategies based on successive backprojections from a desired goal to a set of starting configurations broadened by uncertainty. Brooks [8] described a symbolic "plan checker" that determines if a plan will succeed in the presence of uncertainty.

A recurrent construction in planning is the higher-dimensional space in which parameters of motion or state are orthogonal axes. In *configuration space* (C-space) [36], orthogonal axes are the translational or rotational coordinates of a workpiece. This is the natural space for path planning. Udupa [66] showed that the collision avoidance problem for two objects can be reduced to the problem of moving a point in the vicinity of a "C-space obstacle," which is a convolution of the two objects. The C-space obstacle is closely related to the Minkowski sum of the two objects.

Operation space is formalized by Mason and Brost [45] as a higher-dimensional space describing parameters of operations which may be performed on a given workpiece in a known configuration. It is a space of robot motions, not workpiece configurations. Examples cited in [45] are the fence-pushing spaces and squeeze-grasp diagrams of Brost [12] and the closely related pushing space of Mani and Wilson [40], Kerr and Roth's grasp space [30], and the jamming and wedging diagrams of Siminuvic [64] and Whitney [71].

In Chapter 5 of this work a *configuration map* is developed. A configuration map is a function of two copies of C-space, representing the initial and final configurations of a workpiece (before and after an operation) and mapping into logical values 0 (not possible) and 1 (possible). Configuration maps are used for planning sequences of operations.

Phase space is a well-known construction in which the coordinates and momenta of a particle are the dimensions of the space. Liouville's theorem [31] states that the volume in phase space (representing an ensemble of particles or a probability cloud for one particle) is constant under nondissipative operations. In robotics, where "objects in motion tend to come to rest" [1], Liouville's theorem could only be applied to brief events. (It is included here in the hope that someone will make use of it in robotics.)

1.6. PREVIOUS WORK ON MANIPULATION OF SLIDING WORKPIECES

The subject of this book is directly descended from the work of Mason [44] [42]. This section summarizes some of Mason's results which are used in this work and describes some of the problems approached in [44] which are fully solved here. It also relates the present work to that of Brost [12], of Mani and Wilson [40], and of Erdmann [17].

1.6.1. Sliding as an Important Physical Effect

Mason first identified sliding operations as fundamental to manipulation, and especially to grasping. As argued above, to plan sliding (or more generally, manipulation) operations, it is necessary to study not only the geometric aspects of the interaction between robot and workpiece, but also the physical aspects: the forces generated and the motions those produce.

The prototypical sliding problem is to solve for the motion of a workpiece on a planar surface with friction, when a force is applied to it at a known point. This is a problem in classical mechanics, indeed in quasi-static mechanics. It was recognized but never solved in the heyday of classical mechanics [28] [39] [58], although the answers turn out to be simple and of analytical form. The sliding problem is difficult because the pressure distribution beneath the workpiece is in general unknown. The nineteenth- and early twentieth-century classical mechanicians (cited earlier) assumed a particular form of the pressure distribution, either uniform or with linear variation over the bottom surface of the workpiece, and solved the very difficult mechanics problem which resulted.

Mason realized the only useful result would be one which applied for all pressure distributions, as the pressure distribution is unknown.

1.6.2. Mason's Rules for Rotation [44]

One such result that has already proven its usefulness [40] [12] concerns whether a pushed sliding workpiece will rotate clockwise (CW) or counterclockwise (CCW). This rule may be summarized as follows, referring to Figure 1-3. Construct the friction cone, of half-width $\tan^{-1} \mu_c$, at the point of contact between pusher and workpiece. The friction cone is centered on the normal to the edge, as shown. As will be explained in Section 4.1.1, the line of force applied by the pusher to the workpiece must fall within the friction cone. If the line of force passes to the left (right) of the CM of the workpiece, the workpiece will rotate CW (CCW). If both edges of the friction cone are left (right) of the CM, the line of force must be left (right) of the CM too. If the edges of the friction cone straddle the CM, the line of motion (drawn horizontally in Figure 1-3) acts as a tie-breaker. In Figure 1-3, one edge of the friction cone passes to the left of the CM, one edge passes to the right, and the line of motion passes to the left. Therefore the workpiece will rotate CW.

As this rule makes no mention of the pressure distribution, it is tempting to believe that the rule is understandable purely as a consequence of the symmetries of the lines of force and CM of the workpiece. This is incorrect: there is no symmetry in the problem because the pressure distribution may be asymmetric. The content of the rule is that the asymmetry doesn't matter.

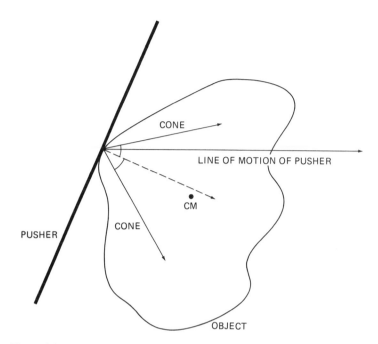

Figure 1-3 Mason's rules for determining CW or CCW rotation of a pushed workpiece. Construct a friction cone of half-width $\tan^{-1} \mu_c$ at the point of contact between pusher and workpiece. The friction cone is centered on the normal to the edge. If the line of force passes to the left (resp. right) of the CM of the workpiece, the workpiece will rotate CW (resp. CCW). The line of force applied by the pusher to the workpiece must fall within the friction cone. If both edges of the friction cone are left (resp. right) of the CM, the line of force must be left (resp. right) of the CM too. If the edges of the friction cone straddle the CM as here, the line of motion acts as a tie-breaker. Here, one edge of the friction cone passes to the left of the CM, one edge passes to the right, and the line of motion passes to the left. Therefore the workpiece will rotate CW.

1.6.3. Center of Rotation

The foregoing rule determines the sense of rotation of a pushed workpiece, but not the rate of rotation. Mason recognized the importance of rate information and framed the problem of determining the center of rotation (COR), which describes the instantaneous motion of a pushed workpiece completely. We cannot hope that the COR will be independent of the pressure distribution, as the sense of rotation is. Rather we may attempt to find the set of CORs as the pressure distribution takes on all functional forms. (This is the problem solved in Chapter 3 of this work.)

Mason's theorem 5 greatly reduces the difficulty of the problem. It states that to find the extremal points of the COR locus (which it is hoped will be a well-behaved compact region of the plane), it is not necessary to consider pressure distributions of *all* functional forms, but only those consisting of three nonzero points of support, called "tripods." Mason states that "while numerical search through all possible three-point pressure distributions is not an appealing prospect, it is certainly more appealing than search through all possible supports" [44].

Even finding the COR for a *single* three-point pressure distribution turned out to be surprisingly difficult, consuming from a few seconds up to a minute of computation on a lisp machine. The technique Mason used was two one-dimensional false-position iterative calculations, corresponding roughly to the two degrees of freedom of the COR. Compared to the two-dimensional energy minimizing iterative method (Section 3.3.3) used in this work to create Figures 3-5 and 3-6, Mason's method is slower but converges reliably. Because of the computational expense, Mason attempted to guess heuristically which tripods provide extremal CORs.

1.6.4. Manipulation Planning and Grasp Planning

Mason [44] also included applications to manipulation planning and to grasp planning. Later work of Mani and Wilson [40] and of Brost [12] dealt more comprehensively with similar applications. Mason analyzed a fence movement across a tabletop which would align and position a small workpiece, in this case a rectangular block, without sensors, by taking advantage of the rotation of the workpiece as it is pushed. The operation was demonstrated and worked with fair reliability. Mani and Wilson extended the idea to general polygonal shapes, and to an arbitrary sequence of pushes of the workpiece with a reorientable movable fence. A special-purpose robot was constructed to demonstrate the idea. Mason also showed how his results could be used to optimize a "push grasp" in which robot fingers close on a workpiece at a controlled rate while the fingers simultaneously translate. The push grasp combines workpiece reorientation with grasping. Brost considered related classes of grasp in more generality and detail.

Besides generalizing the fence push strategy or the push-grasp strategies of Mason, Mani and Wilson [40] and Brost [12] dealt more directly with the issue of planning. The closely related "squeeze-grasp" and "push-grasp" diagrams of Brost [12] and "edge-stability" maps of Mani and Wilson [40] essentially plot Mason's rule for sense of rotation as a function of the parameters of a grasping or manipulation operation. These are then examples of "operation spaces," formalized by Mason and Brost [45]. In [40] and [12] the parameters of an operation are the angle of the fence or gripper surface, and the direction it is translated, relative to the orientation of the workpiece to be grasped or manipulated. Where Mason analyzed the physics of an operation, Mani and Wilson [40] and Brost [12] included the geometry and the physics in one construction. An alternative approach to encapsulating the physics and geometry together is described in Chapter 5 of this work, where configuration maps for the same purpose are introduced.

Erdmann [17] developed a method for manipulation planning based primarily on geometric considerations. The backprojection of a state is the set of configurations from which that state can be achieved despite uncertainty, by executing a particular motion. Plans are formed by chaining backprojections until all possible initial states are included. While primarily geometric, Erdmann's method does include the physics of interaction in that sliding or sticking of workpieces in contact is determined. For the purpose of this determination a representation of the friction cone (Section 4.1.1) generalized to configuration space is introduced.

1.7. SUMMARY OF RESULTS

1.7.1. Chapter 2: The Minimum Power Principle

The motion of a pushed, sliding workpiece, if it is not too fast, is a problem in quasi-static mechanics. In Chapter 2 we study a simple, perhaps obvious, variational principle for quasi-static systems which we call the *minimum power principle*. This principle can be stated

> A quasi-static system chooses that motion, from among all motions satisfying the constraints, which minimizes the instantaneous power.

For our purposes "instantaneous power" may be understood as the rate of energy dissipation to sliding friction. The principle expresses the intuitively appealing idea that when a workpiece is pushed, it gets out of the way of the pusher (i.e., "satisfies the constraints") in the easiest way: the way which minimizes the energy lost to sliding friction.

The relationship of the minimum power principle to other energetic formulations of the laws of mechanics is discussed. Contrary to popular misconception, the minimum power principle is not an existing principle in mechanics, and, in fact, it is false. A simple counterexample is given.

We prove that in the special case of Coulomb friction, the minimum power principle is correct, however. For systems with other dissipative forces (e.g., viscous forces) the minimum power principle produces incorrect results.

1.7.2. Chapter 3: The COR Locus for a Sliding Workpiece

In Chapter 3 we consider the problem of predicting the instantaneous motion of a pushed sliding workpiece. A typical sliding problem is shown in Figure 1-4. The instantaneous motion is completely given by the location of the center of rotation of the workpiece (COR), somewhere in the plane of sliding.

The motion of the workpiece (and therefore the location of the COR) is sensitive to the pressure distribution P which supports the workpiece on the tabletop. P is unknown in practice, so we wish to find the *locus* of all possible CORs, for all P.

Using the minimum power principle, we find analytic expressions describing the boundary of the COR locus. A typical such boundary is shown in Figure 1-6. Depending on the pressure distribution, the COR of the workpiece may fall anywhere within the ice-cream-cone shaped boundary, but never outside of it.

The parameters of the sliding problem are the center of mass (CM) of the pushed workpiece, the point of contact $\vec{c}$ between the pusher and the workpiece, and the angle α between the edge being pushed and the line of pushing, as shown in Figure 1-5. We also require the radius a of a disk circumscribing the workpiece. The COR locus we find is exact for the disk and encloses the COR locus for any inscribed workpiece.

The coefficient of friction μ_s between workpiece and tabletop does not

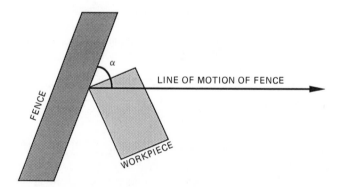

Figure 1-4 A typical pushing problem. A fence moving horizontally, and tipped at an angle α with respect to its line of motion, pushes a corner of a workpiece which is able to slide. The instantaneous motion of the workpiece is completely specified by its *center of rotation*, somewhere in the plane of sliding. The COR cannot be found uniquely because it depends upon unavailable information about the pressure distribution supporting the workpiece. But we will be able to find the COR *locus*, over all pressure distributions.

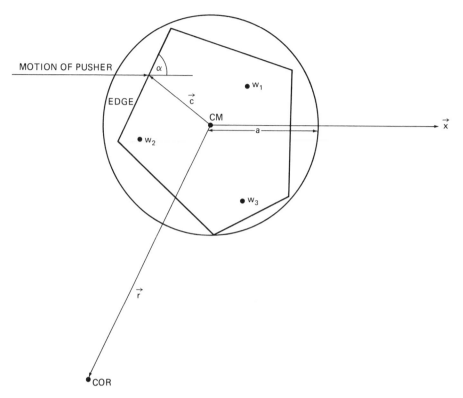

Figure 1-5   Parameters of the pushing problem. Important geometric parameters are the angle α of the edge being pushed relative to the line of motion of the pusher, the vector $\vec{c}$ from the center of mass (CM) to the point of contact between pusher and workpiece, and the radius a of the disk which circumscribes the workpiece. When these parameters are given, the locus of centers of rotation for all possible pressure distributions can be found.

affect the motion so long as we assume uniform Coulomb friction. The coefficient of friction μ_c between the pusher and the edge of the workpiece does affect the motion. In Chapter 3 we assume $\mu_c = 0$.

Note that the COR locus is symmetric about the angle of the edge α, which is drawn as a vector $\vec{\alpha}$ in Figure 1-6. The farthest point of the COR locus from the CM falls on $\vec{\alpha}$. For most applications this "tip" of the COR locus is of particular importance, as it specifies the slowest possible rotation of the workpiece as it is pushed, for any pressure distribution. The distance r_{tip} to the tip of the COR locus has a simple relation to the parameters of the problem:

$$r_{tip} = \frac{a^2}{\vec{\alpha} \cdot \vec{c}}$$

This formula has an interesting geometric interpretation. As the edge angle α is varied, the tip of the COR locus traces out a straight line called the "tip line" and shown in Figure 1-6. The tip line is perpendicular to $\vec{c}$ and a distance a^2/c from the CM. Simple formulas also exist for the curvature of the boundary of the COR locus at the tip (and at the opposite end as well), and for the points of intersection of the boundary of the COR locus with the perimeter of the disk. For most purposes the analytic formulas for these points of the COR locus suffice, and it is unnecessary to generate the entire locus.

As an application of the results so far, we can calculate the maximum distance it is necessary to push a polygonal workpiece with a frictionless fence in order to guarantee alignment of an edge of the workpiece with the fence, regardless of the pressure distribution beneath the workpiece.

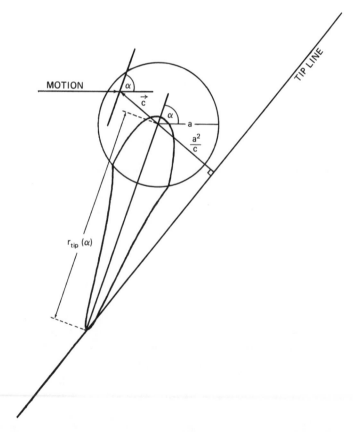

Figure 1-6 $r_{tip}(\alpha)$ **versus** α **and construction of the tip line.** As the angle of the edge being pushed α is varied, the tip of the COR locus boundary traces out a straight line: the tip line.

1.7.3. Chapter 4: The COR Locus Including Contact Friction

The COR loci derived in Chapter 3 apply only when μ_c, the coefficient of friction between the pusher and the pushed workpiece, is zero. In Chapter 4 we generalize to $\mu_c > 0$. The COR locus for $\mu_c > 0$ turns out to be a combination of two of the COR loci calculated for $\mu_c = 0$. The two COR loci used are those with "effective" edge angles $\alpha \pm \tan^{-1} \mu_c$. Part of each of these two loci, plus a linear segment just above the tip line, constitutes all the possible centers of rotation for $\mu_c > 0$. In Figure 1-7 the shaded and bold sections are the resulting COR locus for $\mu_c > 0$.

To demonstrate the use of the results, we find the distance a polygonal workpiece must be pushed by a fence to assure alignment of an edge of the workpiece with the fence, now with $\mu_c > 0$. We also analyze the motion of a

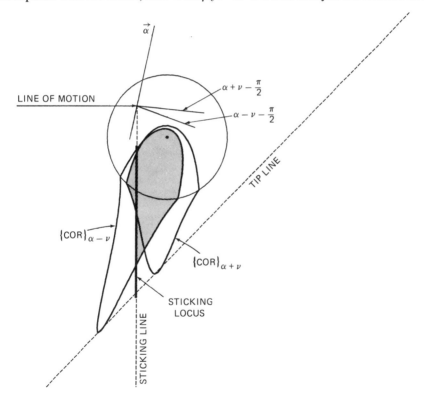

Figure 1-7   Construction of the COR sketch. The COR locus for $\mu_c > 0$ turns out to be a combination of two of the COR loci calculated for $\mu_c = 0$. The two COR loci used are those with "effective" edge angles $\alpha \pm \tan^{-1} \mu_c$. Part of each of these two loci, plus a linear segment just above the tip line, constitutes all the possible centers of rotation for $\mu_c > 0$. The shaded and bold sections are the resulting COR "sketch" for $\mu_c > 0$.

sliding disk as it is pushed aside by the corner of an object in linear motion. Finally, we study the effectiveness of a sensorless manipulation strategy based on "herding" a disk toward a central goal by moving a pusher in a decreasing spiral about the goal.

1.7.4. Chapter 5: Planning Manipulation Strategies

In Chapter 5, a method of planning operation sequences is proposed. The planning problem is to create, given the shape of a workpiece, a sequence of operations which will perform a desired manipulation on the workpiece, despite substantial uncertainty in the initial position and orientation of the workpiece.

We define a *configuration map* which describes the effect of a single operation on all initial configurations of a workpiece. Figure 1-8 shows a configuration map for the interaction of the five-sided workpiece shown with a moving fence tipped at −60 degrees with respect to its direction of motion. The horizontal axis of the configuration map is the initial configuration of the workpiece. The vertical axis is the final orientation of the workpiece after it has collided with the fence, slid along its face, and left the end of the fence. Shaded regions of the map are logical "1," meaning a workpiece with that initial orientation could have the indicated final orientation after interacting with the fence. Both geometric reasoning and the physics of sliding developed here are needed to generate the configuration map. The map does not have a unique final configuration for each initial orientation, because the pressure distribution supporting the workpiece is unknown and affects the motion.

In general a configuration map is a function of two copies of configuration space (C-space × C-space), taking on logical values. In this case we are interested only in the orientation of the workpiece, so we use only a one-dimensional projection of C-space. The utility of configuration maps lies in the fact that the configuration map for a complex sequence of operations is simply the product of configuration maps for the elementary operations composing the sequence.

As an example of planning using configuration maps, we consider a class of passive parts-feeders based on a conveyor belt. Workpieces arrive on the belt in random initial orientations. By interacting with a series of stationary fences angled across the belt, the workpieces are aligned into a unique final orientation independent of their initial orientation. Figure 1-9 illustrates a sequence of six fences which aligns the five-sided workpiece shown. The configuration map for the sequence of six fences has only one horizontal band across it: all initial orientations are reduced to a narrow band of final orientations.

In this example the planning problem is to create (given the shape of a workpiece) a sequence of fences which will align that workpiece. Using

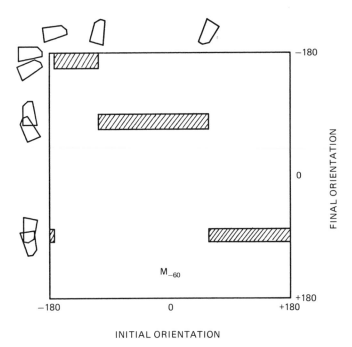

Figure 1-8 Configuration map for workpiece interacting with −60-degree fence.
The *configuration map* describes the effect of a single operation on all initial
configurations of a workpiece. This is the configuration map for the interaction of
the five-sided workpiece shown with a moving fence tipped at −60 degrees with
respect to its direction of motion. The horizontal axis of the configuration map is
the initial configuration of the workpiece. The vertical axis is the final orientation
of the workpiece after it has collided with the fence, slid along its face, and left the
end of the fence. Shaded regions of the map are logical "1," meaning a workpiece
with that initial orientation could have the indicated final orientation after interacting with the fence. The map does not have a unique final configuration for each
initial orientation, because the pressure distribution supporting the workpiece is
unknown and affects the motion.

configuration maps, we transform the planning problem into a purely symbolic one. A tree consisting of all fence sequences is quickly searched to find
a successful feeder design, by using some effective pruning strategies.

1.7.5. Chapter 6: Experiments

In Chapter 6 several experiments are described which test the adherence of real physical systems to the analytic bounds derived in earlier chapters. In general we find that the analytic bounds derived in Chapters 3 and 4
and used in Chapter 5 are rather conservative bounds. But when nonuniform sliding surfaces are used or when lubricants (non-Coulomb friction) are

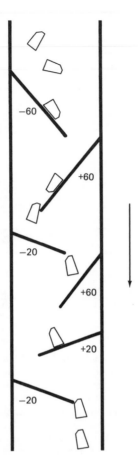

Figure 1-9 Top view of a parts-feeder design. Workpieces arrive on the belt in random initial orientations. By interacting with a series of stationary fences angled across the belt, the workpieces are aligned into a unique final orientation independent of their initial orientation.

present on the surfaces, substantial disagreement with the analytic bounds is expected and observed.

1.7.6. Chapter 7: Suggestions for Further Work

In the final chapter we consider relaxation of some of the assumptions made in earlier chapters, particularly with regard to non-Coulomb friction and high-speed manipulation. We also consider the prospects for extending the configuration maps developed in Chapter 5 to include probability of various results and to incorporate sensory information.

1.8. SIGNIFICANCE OF RESULTS

1.8.1. Physics of Sliding

In Chapters 3 and 4 we find the COR loci which completely specify all possible motions of a sliding workpiece. These results generalize those of Mason [44] which determine the sense of rotation (i.e., whether the workpiece will turn clockwise or counterclockwise). The COR contains *rate* as well as sense information and specifies the motion completely.

There are some real problems which can be solved with sense-of-rotation information alone. In Erdmann and Mason's tray-tipper (Section 1.4), when a workpiece slides into contact with a wall of the tray, we only need to know whether it rotates and which way it rotates to calculate its eventual state. Erdmann's C-space friction cones and geometric reasoning provide this information. But most problems cannot be solved without rate information. In all problems where the pushing motion is of finite length we must calculate the rate of rotation. In Mani and Wilson's fence aligner and Brost's push grasps, it is assumed that a face of the workpiece rotates into alignment with a face of the pusher before the pusher's motion terminates. Rotation rates are needed to calculate the push lengths needed to justify that assumption. The tray-tipper avoids the assumption by not restricting the duration of pusher motion, which is effectively replaced by gravity.

By no means all workpiece-alignment strategies count on alignment with fences. In the conveyor belt–based aligner described above and in Section 5.4, we use the relative rates of workpiece rotation and slippage of a fence's endpoint along a workpiece edge to calculate the eventual orientation of a workpiece after it leaves the endpoint of the fence. In Section 4.5, we find the maximum rate at which a point pusher could "herd" a disk toward a central goal by spiraling around it. Paul's hinge-grasp strategy (Figure 1-2) assumes that the hinge plate will rotate until it contacts both fingers before the converging fingers become too close together to encompass both sleeves. To find the maximum allowable convergence rate requires us to find the minimum possible rotation rate of the hinge plate when it is being pushed by one finger.

1.8.2. Grasp Planning

COR loci are relevant to grasping in two distinct ways:

1.8.2.1. Grasp strength.   When a grasp is formed with a parallel jaw gripper, the maximum torque which the grasp can withstand before slipping depends on where the COR is when that slipping begins. Just as the COR for sliding on a tabletop depends on the details of the contact with the surface, so does the COR for slipping in a gripper. And since in both cases the details are often unknown, the motivation for finding a set of all possible results is the same. Though the grasp strength problem has not been addressed directly in this work, it is closely related.

1.8.2.2. Uncertainty reduction in grasping.   The motion of a pushed sliding workpiece is particularly relevant to grasping actions. Workpieces are often acquired from belts or tabletops on which they could slide.

When grasping is attempted they are invariably pushed, and consequently slide, before being fully constrained by a gripper. It is at this point that carefully maintained knowledge of the workpiece's position and orientation could be lost by bumping it. On the other hand, a haphazardly placed workpiece could be acquired in such a way that knowledge of its position in the gripper is enhanced (e.g., Brost's push grasps [12]). Fundamental to avoiding the former or achieving the latter event is careful consideration of the mechanics of sliding.

A configuration map can be computed for a grasping operation. If the grasping operation consists of several simpler operations, the configuration map for the grasp can be computed as a product of the simpler ones. The product of the configuration map for the grasp with those of preceding and subsequent manipulation operations can be used to keep track of uncertainty, and to plan strategies which control uncertainty.

1.8.3. Assembly Planning

Configuration maps can be used in planning assembly operations, in order to keep track of uncertainty and to represent the elementary operations which may be combined in a motion plan. Unlike grasping, the fully three-dimensional mechanics problems which occur in assembly will only occasionally involve planar sliding. An example of an assembly operation which does involve planar sliding is positioning a washer above a hole prior to dropping a bolt through the washer and hole.

1.8.4. Manipulation Planning

Configuration maps are probably best adapted to the planning of multiple-step manipulation strategies. Parts-feeders are an example of such strategies, but similar strategies for the broader class of motions a robot can perform can be planned by the same method (Section 5.4). Much manipulation can be accomplished before and during grasping, and for these operations planar sliding is a major effect. Because of the limitations of simple grippers, very few manipulation operations can be performed on a workpiece once it is grasped. However, one can imagine intentionally causing the workpiece to rotate or slip in the gripper by pushing the grasped workpiece into a rigid object, or by allowing the workpiece to rotate in the gripper under the influence of gravity (in the way that humans may handle a glass of water). Calculating the effects of these operations requires an understanding of planar sliding.

1.8.5. Parts-Feeders and Robot Manipulation Strategies

Parts-feeders are closely related to sensorless manipulation strategies for robots. The set of elementary operations which can be used in a feeder is somewhat restricted because the unperturbed movement of a workpiece is usually required to be linear (driven by a belt or a vibrating ramp), whereas the motion of a robot may be arbitrary (e.g., the "herding" strategy considered in Section 4.5). On the other hand, the designer of a parts-feeder may consider a broader class of levers, bumps, flaps, and holes for a workpiece to interact with than are available as part of a robot gripper.

The example described in Section 5.4 of a conveyor belt–based parts-feeder, in which the elementary operations are interactions with fixed fences angles across the belt, is usable both as a parts-feeder and as a robot strategy. Indeed, in Mani and Wilson's implementation of a similar system [40], it is the fences that move (as part of a special-purpose robot) and the workpieces that are on a fixed surface.

1.8.6. Practicality of the Belt Parts-Feeder

Considered as a parts-feeder, the conveyor belt–based system described in Section 5.4 may indeed be practical. To implement it, one needs moderate control of the coefficient of friction μ_c between the fences and the workpieces. Of the many designs generated by the algorithm described, for a given workpiece shape, one should be selected for which the sensitivity of the design to μ_c is minimal. Additionally, we shall see in Section 6.6 that nonuniformity of the sliding surface may cause trouble. Here too, the design chosen should be one in which slight shifts in individual configuration maps do not cause the design to fail. The required length of each fence is not addressed in Section 5.4, but can be easily computed from the results of Chapter 4. While speed of operation may be a concern if a robot is to execute a strategy consisting of several large movements, the same concern does not apply for a conveyor belt.

2

The Minimum Power Principle

Quasi-static mechanical systems are those in which mass or acceleration is sufficiently small that the inertial term ma in $F = ma$ is negligible compared to dissipative forces. Many instances of robotic manipulation can be well approximated as quasi-static systems, with the dissipative force being dry friction.

Energetic formulations of Newton's laws have often been found useful in the solution of mechanics problems involving multiple constraints. The following energetic principle for quasi-static systems seems intuitively appealing, or perhaps even obvious:

> A quasi-static system performs the allowed (i.e., unconstrained) motion which minimizes the instantaneous power.

Roughly speaking, the *minimum power principle* states that a system performs the lowest-energy, or "easiest," motion in conformity with the constraints.

Surprisingly, the principle is in general false. For example, if viscous forces act the motion predicted by the minimum power principle will be incorrect. But we prove that the principle is correct in the useful special case that Coulomb friction is the only dissipative or velocity-dependent force acting in the system.

2.1. INTRODUCTION

2.1.1. Quasi-static Systems and the Minimum Power Principle

The quasi-static approximation to the motion of a mechanical system is the solution to Newton's law $F_{total} = ma$ with the inertial term ma ignored. Ignoring ma is only exact in trivial cases, but in many systems dissipative forces so overwhelm the inertial term that the quasi-static approximation is useful.

The quasi-static approximation can be used even when neither mass nor velocity is small. In particular, velocities need not be so small that the motion is only sensitive to the $\nu \to 0$ limit of velocity-dependent forces:

A bacterium swims in a viscous fluid. Dissipative forces are proportional to ν. A bacterium can drift only about 10^{-6} body lengths without swimming [6], so inertial effects are minimal. The shape assumed by the bacterium's flexible flagellum for a given motion at its base can be analyzed in the quasi-static approximation, despite the fact that viscous forces vanish in the $\nu \to 0$ limit.

The quasi-static approximation is appropriate for many interesting driven dissipative systems below a characteristic driving velocity. For systems involving frictional forces, characteristic velocities for quasi-static motion have been discussed in [43] and [54]. Bounds on the error caused by using the quasi-static approximation can be estimated in particular cases.

Example: A credit card on a tabletop, with weight uniformly distributed over the area of contact, rotates as it is pushed by a robot finger. Here we find that a characteristic pushing velocity at which the quasi-static approximation produces 10% errors is roughly 10 cm/sec.

Example: A rope lying snaked on the ground straightens as one end is pulled steadily.

The *minimum power principle* can be stated

A quasi-static system chooses that motion, from among all motions satisfying the constraints, which minimizes the instantaneous power.

For the preceding two examples "instantaneous power" may be understood as the rate of energy dissipation to sliding friction. Note that in each example one of the constraints is a "moving constraint" (one that imposes a motion on the system). Were this not so the systems would choose the lowest power motion of all: no motion.

The minimum power principle expresses the intuitively appealing idea that when the credit card is pushed or the rope is pulled, each "satisfies the constraints" (e.g., gets out of the way of the pushing finger, or complies with the motion of the pulling hand) in the easiest way: the way which minimizes the energy loss to sliding friction.

2.1.2. Relation of the Minimum Power Principle to Other Energetic Principles

Because of its simplicity the minimum power principle seems reminiscent of many other energetic principles in mechanics. This has caused much confusion. The minimum power principle is not an existing principle of mechanics, and in fact, it is false.

In particular the minimum power principle is not related to the well-known "principle of virtual work." The latter, discussed in most elementary mechanics texts, is a method of calculating the internal forces in a system. To find a force exerted by a constraint, one imagines a "virtual displacement" δ of part of the system, violating the constraint. The principle of virtual work states that the change in energy of the system due to δ is equal to δ times the force of constraint. The principle of virtual work is not a method of predicting the motion of the system as the minimum power principle is.

The minimum power principle is also not related to the "principle of least action" in Lagrangian mechanics or to the Hamiltonian formulation of mechanics. These formulations apply only when all forces are conservative, that is, when there are no dissipative forces. (However, see [21] for an extension of Lagrangian mechanics to include radiative and relaxation processes.)

The minimum power principle *is* related to topics in the classical theory of plasticity [62] [13]. These topics are beyond the scope of this chapter. Interested readers may find a summary and further consideration of the validity of the minimum power principle in the work of Goyal and Ruina [22].

All energetic formulations of mechanics, including Lagrangian and Hamiltonian, are ultimately isomorphic to simple Newtonian ($F = ma$) mechanics. In other words, it can be proved that the answers obtained from all formulations are the same. This fact has not diminished the usefulness of the energetic formulations. In dynamic systems, especially with multiple constraints, the energetic principles greatly simplify the solutions because constraint forces need not be evaluated.

The minimum power principle is much less powerful than Lagrangian or Hamiltonian mechanics, dealing as it does only with quasi-static systems. The subject of this chapter may be stated: "Is the minimum power principle isomorphic to Newtonian mechanics in the quasi-static approximation?'' If so, the minimum power principle can be a useful addition to the available techniques for dealing with quasi-static systems. It shares with Lagrangian and Hamiltonian mechanics avoidance of explicit consideration of constraint forces, and it is able to deal with dissipative systems, which the others are not.

In fact we will find that for the isomorphism to hold we must assume not only the quasi-static approximation, but also that all velocity-dependent forces acting on the system act in accordance with the simplest existing model of friction: Coulomb friction. The minimum power principle does not produce correct results for other dissipative forces (e.g., viscous forces or more detailed models of dry friction).

2.1.3. Quasi-static Systems in Robotics

The minimum power principle is used here to solve a problem similar to the earlier "credit card" example. Works of Mason [44], Brost [12], and Mani and Wilson [40] analyze similar scenarios in which a part free to slide on a tabletop is manipulated by a robot. In these works the motion of a sliding object must be determined, which is a problem in quasi-static mechanics.

Trinkle [65] has found the minimum power principle relevant to the planning of robotic grasps in three dimensions.

The dynamics of the robot itself or of its effects on the environment cannot be considered within quasi-static mechanics when kinetic effects are important. However, many other problems arise in robotics which can be partially or completely analyzed in the quasi-static approximation:

The strength and mode of failure of a grasp as external forces are applied to the grasped part.

The stability or mode of collapse of a partially assembled structure.

Prediction of backlash in a system of gears or tendons (with friction, but at low speeds).

The effect of terrain on the trajectory of a mobile robot with coupled wheels, when wheel slip is an issue.

Rigidity (and deviation from nominal shape) of a robot under load, including frictional coupling of the links of the robot. Similarly, rigidity of a part as it is machined under numerical control, and the shape actually cut.

2.1.4. Constraints

In testing the correctness of the minimum power principle we compare its solution to that of Newton's law. We are interested in *n*-particle systems including multiple constraints, so the treatment of those constraints is important. Constraints enter the minimum power principle solution only indirectly, as a limitation on the space of motions over which instantaneous power is minimized. However, the forces which maintain the constraints must be considered explicitly in the Newtonian solution.

To compare the solutions we introduce $3n$-dimensional *constrained directions*, which mesh neatly with the method of Lagrange multipliers in the Newtonian solution. In the minimum power principle solution, the same constrained directions are the basis vectors of a subspace complementary to that over which instantaneous power is minimized.

Constraints are central to the analysis of the foregoing example systems. In the "rope" example, the rope, which is a continuous object, may be approximated by an arbitrarily dense linear collection of point particles, each constrained to be at a fixed (small) distance from its two adjacent neighbors.

The credit card may be considered to be a network of point particles, each constrained to lie at fixed distances from several nearby particles. With enough such constraints the object is rigid. The credit card and the rope are also affected by an external constraint that keeps them in the plane of the tabletop or of the ground, respectively. And each system is affected by an external, moving constraint: the robot finger or the hand pulling the rope.

Of course one would not normally analyze a rigid object as a collection of particles and constraints. Simpler specifications of it are possible, having as few as six degrees of freedom and no internal constraints. We will use the "collection of particles" specification in discussing the validity of the minimum power principle, because that specification is completely general. In actually *using* the minimum power principle, simpler specifications would be employed. This issue is discussed further in Section 2.5.

2.1.5. What Is a Constraint?

Real forces exerted on a particle are always continuous functions of the particle's position. The forces of constraint mentioned earlier are so abrupt, however, that a useful idealization is to consider them to be due to perfectly rigid links, enforcing fixed distances. This idealization is useful because with sufficient rigidity the detailed nature of the forces is unimportant to the motion. However, the idealization brings with it difficulties in calculation due to the singularities which may arise.

We therefore segregate the forces which act in a system into two classes. One class, which we will call $\vec{F}_C$, consists of forces due to the idealized rigid constraints. The second class contains all remaining forces, and will be denoted $\vec{F}_{XC}$. (XC stands for "except constraints.") $\vec{F}_{XC}$ may include external fields (e.g., gravitational, electric, magnetic), dissipative forces (e.g., friction, viscosity), and interparticle forces (e.g., spring forces). We have $\vec{F}_{total} = \vec{F}_C + \vec{F}_{XC}$. Newton's law is simply $\vec{F}_{total} = 0$ in the quasi-static approximation.

2.1.6. The Minimum Power Principle

We define the "instantaneous power" P_ν of a system of particles to be

$$P_\nu = - \sum_i \vec{F}_{XC_i} \cdot \vec{v}_i \qquad (2.1)$$

where i ranges over the particles, $\vec{F}_{XC_i}$ is all forces acting on particle i except forces of constraint, and $\vec{v}_i$ is the velocity of particle i.

Dissipative forces (such as friction) contribute negatively to this sum, and conservative forces can contribute with either sign. Constraint forces, including moving constraints, do not contribute to P_ν. Because forces of constraint are left out of $\vec{F}_{XC}$, P_ν bears no obvious relation to actual energies of the system.

As P_ν is insensitive to mass and acceleration, the minimum power principle cannot give the correct result (i.e., the one which agrees with Newton's law) for non-quasi-static systems. Our purpose is to find out whether the minimum power principle gives the correct result for quasi-static systems. The minimum power principle is not in general isomorphic to Newton's law even for quasi-static systems, and an example of their disagreement is given in Section 2.5. We will find that a sufficient condition for isomorphism is that all velocity-dependent forces acting in the system must be essentially equivalent to Coulomb friction. (All dissipative forces, and some conservative forces, are velocity dependent.)

We will be considering two ways of finding the motion of a system involving constraints without becoming entangled with singularities. The first of these is the familiar method of Lagrange multipliers, using Newton's law. In the second we use the minimum power principle. By equating the two solutions, we determine restrictions on the types of forces for which the minimum power principle gives the correct (i.e., consistent with Newton's law) solution.

We will first consider a single-particle system without constraints. A few lines of vector algebra are sufficient to find the restrictions on the types of forces. In Section 2.3 we introduce constraints in terms of "constrained directions" along which the projection of velocity must be zero. In Section 2.4 we generalize the forces from three dimensions to $3n$ dimensions to represent an n-particle system. The constrained directions generalize easily to $3n$ dimensions. The equations derived for the one-particle case retain their form when generalized to n-particles. Finally we consider a simple example.

2.2. ONE-PARTICLE SYSTEMS WITHOUT CONSTRAINTS

We will assume that the system has arrived at its present state in accordance with the laws of physics and ask only what happens in the next moment. The instantaneous velocity alone completely answers that question. We can find the instantaneous velocity by using Newton's law or the minimum power principle.

2.2.1. Newton's Law

The Newtonian solution for the instantaneous velocity of a particle in the quasi-static approximation is that velocity that satisfies

$$\vec{F}_{total} = 0 \tag{2.2}$$

In the absence of constraints, $\vec{F}_{XC} \equiv \vec{F}_{total}$.

2.2.2. Minimum Power Principle

With P_ν as defined in equation (2.1), and in the absence of constraints, the velocity specified by the minimum power principle is the one for which

$$\nabla P_\nu = 0 \tag{2.3}$$

Note that the gradient is taken with respect to $\vec{\nu}$, the possible motions.

If we had constraints, they would enter equation (2.3) only as a restriction on the vector space over which P_ν is minimized.

2.2.3. Equating the Solutions

We wish to find the conditions under which equations (2.2) and (2.3) are satisfied for the same velocity $\vec{\nu}$, that is, where the minimum power principle gives the same solution as Newton's law.

A necessary and sufficient condition for equivalence of the solutions is that the left side of equation (2.2) is zero exactly where the left side of equation (2.3) is zero. We will use the stronger (sufficient) condition that the left sides are equal. Equating the left sides of equations (2.2) and (2.3) and using the definition of P_ν from equation (2.1), we have

$$\vec{F}_{total} = \nabla(\vec{F}_{XC} \cdot \vec{\nu}) \tag{2.4}$$

Since $\vec{F}_{XC} \equiv \vec{F}_{total}$ we now drop the subscripts. Equation (2.4) may be broken into scalar components and transformed:

$$\forall_j \quad F_j = \frac{d}{d\nu_j}(\vec{F} \cdot \vec{\nu}) \tag{2.5}$$

$$\forall_j \quad F_j = \frac{d}{dv_j} \sum_i F_i v_i$$

$$\forall_j \quad F_j = \sum_i v_i \frac{d}{dv_j} F_i + \sum_i F_i \frac{dv_i}{dv_j}$$

$$\forall_j \quad F_j = \sum_i v_i \frac{d}{dv_j} F_i + F_j$$

$$\forall_j \quad 0 = \sum_i v_i \frac{d}{dv_j} F_i \qquad (2.6)$$

The indices i and j run from 1 to 3, as we are dealing with one particle in 3-space. In later sections we will generalize to n particles in $3n$-space, with i and j running from 1 to $3n$.

Equation (2.5) [or (2.6)] is a sufficient condition, in its most general form, on the types of forces for which the minimum power principle gives the correct solution.

2.2.4. Forces for Which the Minimum Power Principle Is Correct

Note that equation (2.5) is linear. If two types of forces individually satisfy (2.5), their sum will also.

If a force is independent of velocity, its derivative with respect to any component of velocity will be zero, so it will satisfy (2.6). Therefore the minimum power principle is valid for all velocity-independent forces. Most common external forces (electric fields, springs, gravity) are velocity independent. A magnetic field acting on a moving electric charge, however, exerts a velocity-dependent force.

If a force $\vec{F}$ is perpendicular to $\vec{v}$, $(\vec{F} \cdot \vec{v})$ in equation (2.5) is zero. Therefore equation (2.5) cannot be satisfied. The minimum power principle does not find the correct solution for forces which are perpendicular to the velocity which gives rise to them. A magnetic field acting on a moving electric charge is an example of a perpendicular force. This result is not surprising: a perpendicular force can do no work on a particle, and so is invisible in P_v. Yet it does affect the motion.

Finally, consider forces which are parallel to the velocity which gives rise to them. We may write

$$\vec{F} = F\bar{v} \qquad (2.7)$$

where F is a scalar and $\bar{v}$ is a unit vector in the direction of $\vec{v}$. Condition (2.4) becomes

$$F\bar{v} = \nabla(F|v|) \qquad (2.8)$$

$$F\bar{\nu} = |\nu|\nabla F + F\nabla|\nu|$$

$$F\bar{\nu} = |\nu|\nabla F + F\bar{\nu}$$

$$0 = |\nu|\nabla F \tag{2.9}$$

To satisfy equation (2.9), the gradient of F must be zero. Therefore $\vec{F}$ must be a constant. Such forces are generalized versions of Coulomb friction, where the frictional force is directed opposite to the velocity, but the magnitude of that force is independent of velocity and direction.

We can conclude that the minimum power principle does not in general give the correct solution for the motion of a one-particle quasi-static system. However, the solution is correct if the forces acting can be composed of

- Forces independent of the velocity of the particles.
- Forces parallel to the velocity of the particles, but whose magnitude is independent of the velocity of the particles.

2.3. ONE-PARTICLE SYSTEMS WITH CONSTRAINTS

In this section we include constraints in the Newtonian and minimum power principle solutions for the motion of a system. By formulating both solutions in terms of the same "constrained directions" along which the projection of the particle's velocity must be zero, the constraint forces in the two solutions are shown to cancel exactly. The question of the equivalence of the Newtonian and minimum power principle solutions is thus reduced to the previous case in which no constraints were involved. The constrained directions will be generalized in Section 2.4 to $3n$ dimensions.

2.3.1. Newtonian Solution by Lagrange Multipliers

When there is a constraint, there is a force to maintain the constraint. These "forces of constraint" must be included in $\vec{F}_{total} = 0$. Generally, the forces of constraint are unknown and cannot be solved directly. The method of *Lagrange multipliers* [20] has been developed to deal with constraints.

In a formulation of the method of Lagrange multipliers well suited to our purposes, each constraint is replaced by a spring that exerts a force proportional to the difference between its length and its "relaxed" length d. We denote the proportionality constant λ. As $\lambda \to \infty$, the spring becomes rigid and therefore acts as a constraint. Recall that rigid constraints were themselves only idealizations of real forces so sharp that their details ceased to be relevant to the motion of a system. Therefore the choice of a very stiff spring to replace the constraint does not reduce the generality of the constraints.

The force exerted by a spring with spring constant λ constraining a particle to be a distance d from the origin is

$$\vec{f}_s = \lambda(d - |\vec{r}|)\bar{r} \qquad (2.10)$$

where $\vec{r}$ is the position of the particle and $\bar{r}$ indicates a unit vector in the direction of $\vec{r}$.

We have initially a state of the system (described by the vector $\vec{r}$) which satisfies the constraints, and ask what happens in the next instant dt. We wish to find $\vec{v}$, the vector specifying the instantaneous velocity of the particle. If a particle is constrained to be a distance d from the origin, and is presently at that distance, then the constraint may be stated as a restriction on the instantaneous velocity of the particle: $\vec{v}$ must be perpendicular to $\vec{r}$. The force arising from a violation of this constraint is

$$\vec{f}_s = -\lambda(\vec{v} \cdot \bar{r})dt \, \bar{r} \qquad (2.11)$$

$\bar{r}$ here is a *constrained direction:* the velocity must be perpendicular to this direction. Figure 2-1 illustrates the constrained direction $\bar{r}$. The velocity of

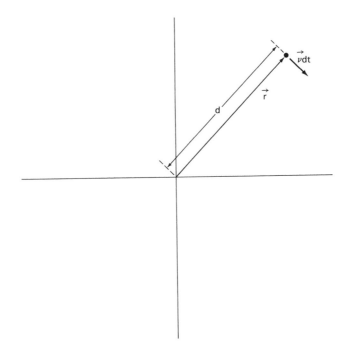

Figure 2-1 $\vec{r}$ **is a constrained direction.** If the particle (dot) is constrained (by a rope, perhaps) to lie a fixed distance d from the origin, then the vector $\vec{r}$ is a constrained direction for the particle. This means that the particle's instantaneous velocity $\vec{v}$ must have no component in the direction $\bar{r}$.

the particle $\vec{v}$, if it is not to violate the constraint, must be perpendicular to the constrained direction. Should it not be perpendicular, the distance from the origin to the particle would increase by $\vec{v} \cdot \vec{r}\, dt$, and a force of constraint $\vec{f}_s$ would develop as given by equation (2.11).

We will require two general forms for forces of constraint. The first,

$$\vec{f}_s = -\lambda(\vec{v} \cdot \bar{c})dt\, \bar{c} \qquad (2.12)$$

is used to enforce a fixed distance from a particle to a point in space. (It can be used for fixed interparticle distances, too, as we will see in Section 2.4.) By properly selecting constrained directions $\bar{c}$ in velocity space, equation (2.12) is sufficient to represent general distance constraints. Suppose a particle at $\vec{r}$ is constrained to lie a distance d from a point $\vec{p}$ fixed in space. Its velocity $\vec{v}$ must be perpendicular to $(\vec{r} - \vec{p})$. The unnormalized vector $\vec{c}$ which represents this constraint is

$$\vec{c}_x = \vec{r}_x - \vec{p}_x$$
$$\vec{c}_y = \vec{r}_y - \vec{p}_y \qquad (2.13)$$
$$\vec{c}_z = \vec{r}_z - \vec{p}_z$$

The second form of constraint we shall need imposes a velocity $\vec{e}$ on the motion of a particle. We can write a force term to maintain this constraint

$$\vec{f}_s = -\gamma[(\vec{v} - \vec{e}) \cdot \bar{e}]\bar{e}\, dt \qquad (2.14)$$

where γ is another spring constant like λ. This equation may be interpreted to say that if the component of the particle's velocity $\vec{v}$ in the $\bar{e}$ direction differs from $\vec{e}$, we will impose a force in the $\bar{e}$ direction.

Newton's law may now be written as

$$\vec{F}_{XC} - \sum_j \lambda_j(\vec{v} \cdot \bar{c}_j)\bar{c}_j\, dt - \sum_k \gamma_k[(\vec{v} - \vec{e}_k) \cdot \bar{e}_k]\bar{e}_k\, dt = 0 \qquad (2.15)$$

where the second and third terms are the forces of constraint from equations (2.11) and (2.13). $\vec{F}_{XC}$ represents all forces other than the constraints.

To solve the system, one must solve for the components of $\vec{v}$ in terms of the multipliers λ_j and γ_k, and then take the limit as all the multipliers go to infinity.

2.3.2. Minimum Power Principle Solution with Constrained Directions

A quasi-static system chooses that motion, from among all motions satisfying the constraints, which minimizes P_v.

In the notation developed, P_v may be written

$$P_v = -\vec{F}_{XC} \cdot \vec{v} \qquad (2.16)$$

$\vec{F}_{XC}$ represents all forces other than the constraints. P_ν is a scalar quantity, while $\vec{F}_{XC}$ and $\vec{\nu}$ are vectors. Were it not for the restriction "among all motions satisfying the constraints," the motion minimizing P_ν would satisfy

$$\nabla P_\nu = 0 \qquad (2.17)$$

If certain directions of motion $\vec{s}_l$ violate the constraints, we do not care if P_ν could be further lowered by moving in those directions. So we only require that P_ν is at a minimum when we change $\vec{\nu}$ in unconstrained directions. In terms of the gradient of P_ν, we do not insist that it be zero in all directions, but only in the unconstrained directions. In the constrained directions the gradient of P_ν may be nonzero. This requirement may be written

$$\nabla P_\nu = \sum_l \alpha_l \vec{s}_l \qquad (2.18)$$

Note that the minimum power principle is satisfied if equation (2.18) is true for any set of values of the parameters α_k. Another way of understanding this is that we require P_ν to be minimized not over the entire velocity space (of dimension 3 now, but which will be generalized to $3n$), but only on a subspace reduced in dimensionality by the number of constraints. The basis vectors of this subspace are perpendicular to all the constrained directions $\vec{s}_l$. P_ν is also defined on the complementary subspace whose basis vectors are the constrained directions $\vec{s}_l$, but it is of no interest what the projection of ∇P_ν onto this space is, because the system is constrained to have zero velocity in this subspace. The minimum power principle therefore allows ∇P_ν to be composed of an arbitrary linear combination of the constrained directions.

2.3.3. Forces for Which the Minimum Power Principle Is Correct

We now wish to find the conditions under which equations (2.15) and (2.18) are satisfied for the same velocity $\vec{\nu}$, that is, where the minimum power principle gives the same solution as Newton's law. When that occurs we have

$$\vec{F}_{XC} - \sum_j \lambda_j(\vec{\nu} \cdot \vec{c}_j)\vec{c}_j \, dt - \sum_k \gamma_k[(\vec{\nu} - \vec{e}_k) \cdot \vec{e}_k]\vec{e}_k \, dt$$
$$= \nabla(\vec{F}_{XC} \cdot \vec{\nu}) - \sum_l \alpha_l \vec{s}_l \qquad (2.19)$$

The constrained directions $\vec{s}_l$ in the minimum power principle solution are the directions along which the projection of velocity must be zero to satisfy the constraints. That is also what the vectors $\vec{c}_j$ and $\vec{e}_k$ are, in the Newtonian solution. The $\vec{s}_l$ are simply a relabeling of the $\vec{c}_j$ and the $\vec{e}_k$. The values α_l

may be chosen arbitrarily, so we choose α_l to be of the form

$$\alpha = \lambda(\vec{v} \cdot \vec{c}) \, dt \tag{2.20}$$

or

$$\alpha = \gamma[(\vec{v} - \vec{e}) \cdot \vec{e}] \, dt$$

depending on whether $\vec{s}_l$ corresponds to a $\vec{c}_j$ or a $\vec{e}_k$. Then the summations in equation (2.19) cancel, leaving only

$$\vec{F}_{XC} = \nabla(\vec{F}_{XC} \cdot \vec{v}) \tag{2.21}$$

The algebra of equations (2.5) to (2.6) applies directly to this equation. The conclusions of Section 2.2.4 therefore apply to one-particle systems with constraints, as well.

2.4. *n*-PARTICLE SYSTEMS WITH CONSTRAINTS

We now generalize the foregoing results to n-particle systems. Both the algebra in Section 2.2 and the constrained directions in Section 2.3 generalize to the $3n$-dimensional velocity space needed for n particles. Henceforth, all vectors will be assumed to be $3n$ dimensional. c_{ix} will denote the x component of $\vec{c}$ for particle i. If a vector is only three dimensional, it will be indicated as, for example, ${}^3\vec{c}$.

2.4.1. Newtonian Solution by Lagrange Multipliers

If a system consists of n particles, we can consider a force $\vec{F}$ to be a vector of $3n$ components. $\vec{F}_{total} = 0$ then describes the Newtonian solution for the whole system at once.

Equation (2.12) gave the force required to maintain a constrained direction $\vec{c}$. It is merely a formality to translate equation (2.12) to a general constrained direction in $3n$-space,

$$\vec{f}_s = -\lambda(\vec{v} \cdot \vec{c}) \, dt \, \bar{c} \tag{2.22}$$

where $\vec{c}$ is a $3n$-vector with components

$$c_{ix} = {}^3c_x$$
$$c_{iy} = {}^3c_y \tag{2.23}$$
$$c_{iz} = {}^3c_z$$

and i is the particle number of the constrained particle. The other $3n - 3$ components of $\vec{c}$ are zero.

The force required to impose a velocity $^3\vec{e}$ on the motion of a particle [equation (2.14)] also generalizes trivially. We can write a force term to maintain this constraint

$$\vec{f_s} = -\gamma[(\vec{v} - \vec{e}) \cdot \vec{e}]\vec{e}\, dt \tag{2.24}$$

where

$$e_{ix} = {}^3e_x$$
$$e_{iy} = {}^3e_y \tag{2.25}$$
$$e_{iz} = {}^3e_z$$

The other $3n - 3$ components of $\vec{e}$ are zero.

Many other constraints relating movement of the particles, most of them unphysical, can be expressed in terms of $3n$-vector "constrained directions" $\vec{c}$ or $\vec{e}$. In an *n*-particle system we need a constraint maintaining interparticle distances. The $3n$-vector $\vec{c}$ which represents a constrained direction for an interparticle constraint has six nonzero components. If particle p has position $^3\vec{p}$, and particle q has position $^3\vec{q}$, then the $3n$-vector $\vec{c}$ which constrains them to maintain their current distance is

$$\vec{c}_{px} = {}^3\vec{p}_x - {}^3\vec{q}_x$$
$$\vec{c}_{py} = {}^3\vec{p}_y - {}^3\vec{q}_y$$
$$\vec{c}_{pz} = {}^3\vec{p}_z - {}^3\vec{q}_z$$
$$\vec{c}_{qx} = {}^3\vec{q}_x - {}^3\vec{p}_x \tag{2.26}$$
$$\vec{c}_{qy} = {}^3\vec{q}_y - {}^3\vec{p}_y$$
$$\vec{c}_{qz} = {}^3\vec{q}_z - {}^3\vec{p}_z$$

The other $3n - 6$ components of $\vec{c}$ are zero. These three types of constraints [equations (2.23), (2.25), and (2.26)] allow us to tie a particle to a given point in space by a fixed-length link, to impose a velocity on a particle, and to tie two particles of a system to each other by a fixed-length link. Thus rigid bodies may be modeled by specifying three non-coplanar interparticle constraints for each particle. Nonrigid bodies (e.g., a rope) may be modeled by specifying fewer constraints (two per particle in the case of a rope, as discussed in Section 2.1.4).

All the equations in the preceding sections apply to $3n$-dimensional velocities as well as to the 3-dimensional velocities for which they were explained. Therefore we can generalize the conclusion of Section 2.2 to apply to *n*-particle systems as well: the minimum power principle does not in general give the correct solution for the motion of a quasi-static system. However, the solution is correct if the forces acting can be composed of

- Forces of constraint.
- Forces independent of the velocity of the particles.
- Forces parallel to the velocity of the particles, but whose magnitude is independent of the velocity of the particles. Such forces are essentially Coulomb friction.

2.5. EXAMPLES

Note that in using the minimum power principle, it is not necessary to model the problem as a collection of particles and constraints. That was done only for purposes of generality in the foregoing sections. Any set of parameters which includes all the degrees of freedom of the system may be used. The required constraints are only those which impose restrictions on the parameters chosen.

For instance, in Section 2.1.4, we mentioned a system in which a credit card slides on a tabletop. The card can be considered to be a network of point particles connected by so many constraints that the network becomes rigid. But the minimum power principle can also be applied to a much simpler specification of the card: we may consider only the 3-space coordinates of 3 non-colinear points of the card. In that case the only constraints which are needed are those which constrain the three points to lie in the plane of the tabletop, and the moving constraint which forces it to move. A still simpler specification of the card is one in which only the x and y coordinates of one point of the card are used, with the z coordinate understood to be that of the tabletop. One angle describing the orientation of the card must also be given. In this specification no constraints besides the moving constraint are needed.

The minimum power principle becomes most advantageous when there are numerous constraints. However, we can demonstrate its use on a very simple system. As an example, consider the two-dimensional one-particle system shown in Figure 2-2. A moving constraint imposes a velocity v_x on the particle, in the $+x$ direction. (The constraint could be a frictionless vertical fence.) The constraint applies a force only in the $+x$ direction. An external constant force (e.g., gravity) acts in the $-y$ direction with magnitude mg. A dissipative force ηv^n opposes the velocity v of the particle. (η should not be interpreted as a coefficient of friction, as we have not defined any normal force which gives rise to it. In particular, note that "gravity" acts in the y direction rather than perpendicular to the plane of motion.)

Coulomb friction corresponds to $n = 0$, viscous friction to $n = 1$. If $n = 0$ and $\eta < mg$, the particle will accelerate in the $-y$ direction, violating the quasi-static approximation, so we will assume $\eta > mg$. After motion begins, the particle will approach a terminal velocity. Until the terminal

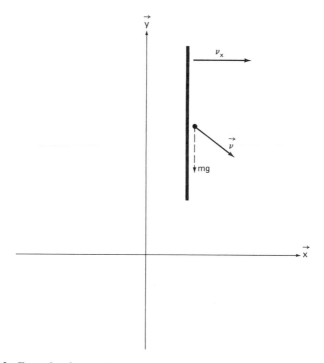

Figure 2-2 Example of a quasi-static system. A moving constraint imposes a velocity v_x on the particle, in the $+x$ direction. (The constraint could be a frictionless vertical fence.) The constraint applies a force only in the $+x$ direction. An external constant force (e.g., gravity) acts in the $-y$ direction with magnitude mg. A dissipative force ηv^n opposes the velocity v of the particle. We wish to solve for the instantaneous velocity $\vec{v}$ of the particle.

velocity is achieved, the motion of the particle is sensitive to its mass, so the quasi-static approximation is not appropriate. We will consider only the time period after inertial effects have been damped out. Motion will then be uniform with time. We wish to find the velocity v_y of the particle as a function of v_x and the dissipative parameters η and n.

2.5.1. Newtonian Solution

The external force mg must be equal to the y component of the dissipative force:

$$mg = f_y = \eta v^n \frac{v_y}{v} \tag{2.27}$$

The constraint moving at velocity v_x determines the x component of the particle's velocity. Using

$$v^2 = (v_x^2 + v_y^2) \tag{2.28}$$

we obtain an implicit solution for v_y:

$$\frac{mg}{\eta} = v_y(v_x^2 + v_y^2)^{(n-1)/2} \tag{2.29}$$

2.5.2. Minimum Power Principle Solution

Instantaneous power due to the external force is $-mgv_y$. The dissipative force is ηv^n, so power is ηv^{n+1}. Total power is then

$$P_\nu = -mgv_y + \eta(v_y^2 + v_x^2)^{(n+1)/2} \tag{2.30}$$

v_x is constrained; v_y unconstrained. We minimize P_ν with respect to v_y:

$$0 = \frac{dP_\nu}{dv_y} = -mg + \frac{n+1}{2}\eta(v_y^2 + v_x^2)^{(n-1)/2}2v_y \tag{2.31}$$

Solving we find

$$\frac{mg}{\eta} = v_y(v_x^2 + v_y^2)^{(n-1)/2}(n+1) \tag{2.32}$$

which is equivalent to the correct answer [equation (2.29)] only when $n = 0$. As concluded in Section 2.2.4, for the minimum power principle to be correct, dissipative forces must be velocity independent, that is, $n = 0$.

3

The COR Locus for a Sliding Workpiece

In this chapter we consider the motion of a workpiece free to slide on a surface. Physical analysis of the workpiece's motion is made difficult by the absence of information about the pressure distribution of the workpiece and of the resulting frictional forces.

The instantaneous motion of the workpiece can be described as a pure rotation about a center of rotation (COR) somewhere in the plane. We find the locus of CORs for all possible pressure distributions, given only the point of application and the direction of a pushing force.

In one application to robotic manipulation, bounds on the distance a workpiece must be pushed to come into alignment with a frictionless robot finger or a fence are determined.

3.1. RANGE OF APPLICABILITY

3.1.1. Workpiece Shape

In this chapter we will treat the workpiece as a two-dimensional rigid body, since we are only concerned with the interaction of the workpiece with the table on which it is sliding. All pushing forces will be restricted to lie in the plane of the table. The results may be applied to three-dimensional workpieces, so long as the vertical component of the pushing force is negligible, and so long as the point of contact is near the table.

3.1.2. Point of Contact Between Workpiece and Pusher

In the general case, when a workpiece is being pushed, there is only one point of contact between the workpiece and the pusher. The contact may be where the flat edge of a pushing fence or robot finger touches a corner of the workpiece (Figure 3-1), or it may be where a pushing point touches an edge of the workpiece (Figure 3-2). In most of this chapter we will assume that the pusher is a point in contact with a flat facet of the workpiece, but the analysis applies equally well if the pusher is a flat surface in contact with a corner of the workpiece.

Motion of a workpiece when there are two or more points of contact between pusher and workpiece has been considered by Brost [12] and by Mani and Wilson [40].

3.1.3. Position Controlled Pusher

It is assumed the pusher will move along a predetermined path in the plane, that is, it is under position control. Equivalently, the surface on which the workpiece slides may move, carrying the workpiece relative to a fixed pusher, for example, on a conveyor belt. The workpiece has two degrees of freedom, with the third degree of freedom of its motion fixed by the contact maintained between the pusher and the workpiece. Our results may be easily converted to the case where the pusher exerts a known force on the workpiece rather than following a known path.

3.1.4. Center of Rotation (COR)

The two degrees of freedom of the workpiece are most conveniently expressed as the coordinates of a point in the plane called the center of

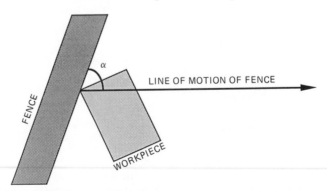

Figure 3-1   The edge of an advancing fence pushing a corner of a sliding workpiece. The motion of the workpiece depends on the angle (α) of the front edge of the fence, measured relative to its line of motion, which in this case is horizontal.

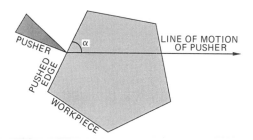

Figure 3-2   A corner of an advancing pusher pushing an edge of a sliding workpiece. The motion of the workpiece depends on the angle (α) of the edge being pushed, measured relative to the line of motion of the pusher, which in this case is horizontal. Compare to the meaning of α in Figure 3-1. The analysis done in this chapter applies equally well to either figure.

rotation. Any infinitesimal motion of the workpiece can be expressed as a rotation $d\theta$ about some COR, chosen so that the infinitesimal motion of each point $\vec{w}$ of the workpiece is perpendicular to the vector from the COR to the point $\vec{w}$. If the workpiece is a disk, and the motion it performs is pure rotation in place, the COR is at the center of the disk. Motions we might describe as "mostly translation" correspond to CORs far from the point of contact. In the extreme case, pure translation occurs when the COR is at infinity.

All kinematic results can be obtained once the COR is found.

3.1.5. Pressure Distribution Between Workpiece and Table

The weight of a workpiece is supported by a collection of contact points between the workpiece and the table. The pressure distribution may change as the workpiece moves relative to the table. Finding the COR is complicated by the fact that changes in the pressure distribution under the workpiece substantially affect the motion; that is, such changes affect the location of the COR. Intuitively, if pressure is concentrated near the center of mass (CM), the workpiece will tend to rotate more and translate less than if the pressure is uniformly distributed over the entire bottom surface of the workpiece.

The pressure distribution may be changed dramatically by tiny deviations from flatness in the workpiece's bottom surface (or of the surface it is sliding on). Indeed, if the workpiece and the table are sufficiently rigid and not perfectly flat, they may be expected to make contact at only three points. The three points may be located anywhere on the workpiece's bottom surface, but like the legs of a three-legged stool, the triangle formed by the points of support always encloses the projection of the CM onto the surface.

Since any assumption we could make about the form of the pressure distribution (for instance, that it is uniform under the workpiece as in [58]) would not be justified in practice, our goal is to find the locus of CORs under *all* possible pressure distributions.

Let the CM be at the origin, and $\vec{w}$ be a point in the plane. All that is known about the pressure distribution $P(\vec{w})$ is that

- $P(\vec{w})$ is zero outside the workpiece. The workpiece can be entirely contained within a circle of radius a centered at the CM.
- $P(\vec{w}) \geq 0$ everywhere.
- the total pressure $\int P(\vec{w})d\vec{w} = Mg$, the weight of the workpiece.
- the first moment of the distribution $\int P(\vec{w})\vec{w}\, d\vec{w} = 0$. This means that the centroid of the distribution is at the CM of the workpiece, which is at the origin.

3.1.6. Coulomb Friction

It turns out that the coefficient of friction of the workpiece with the supporting surface (called μ_s for "sliding friction") does not affect the motion of the workpiece if we use a simple model of friction. We assume that μ_s is constant over the work surface, that it is independent of normal force magnitude and tangential force magnitude and direction (isotropic), and that it is velocity independent. In short, we assume Coulomb friction.

There is another coefficient of friction in the problem, μ_c (for "contact friction"), at the point of contact between the edge of the workpiece and pusher. This is distinct from the coefficient μ_s between workpiece and table, discussed earlier. Initially we consider only $\mu_c = 0$. This assumption is relaxed in Section 4.1.

3.1.7. Quasi-static Motion

It is assumed that all motions are slow. This *quasi-static approximation* requires that frictional forces on the workpiece (due to the coefficient of friction with the surface μ_s) quickly dissipate any kinetic energy of the workpiece:

$$v^2 \ll Xg\mu_s \qquad (3.1)$$

where v is the velocity of the workpiece, g is the acceleration due to gravity, and X is the precision with which it is desired to calculate distances. The high-speed limit is discussed in Section 7.4. Characteristic speeds for quasi-static motion are discussed in [54] and [43].

3.1.8. Bounding the Workpiece by a Disk

We will take the workpiece being pushed to be a disk with its CM at the center. Given another workpiece of interest we can consider a disk centered at the CM of the workpiece, big enough to enclose it. The radius a of the disk is the maximum distance (from the CM) of the workpiece to any point of the workpiece. Since any pressure distribution on the workpiece could also be a pressure distribution on the disk, the COR locus of the disk must enclose the COR locus of the workpiece. The locus for the disk provides useful bounds on the locus for the real workpiece.

3.1.9. Geometric Parameters

Geometric parameters of the problem are the point of contact $\vec{c}$ between the pusher and the workpiece and the angle α between the edge being pushed and the line of pushing, as shown in Figure 3-3. The values of α and

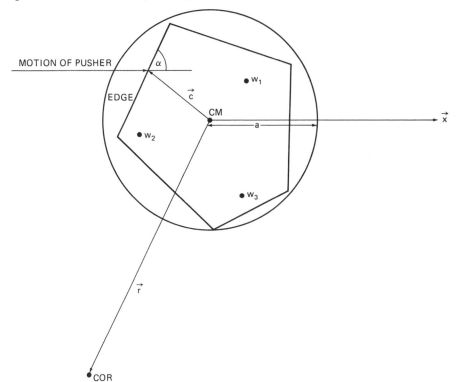

Figure 3-3   Parameters of the pushing problem. Important geometric parameters are the angle α of the edge being pushed relative to the line of motion of the pusher, the vector $\vec{c}$ from the center of mass (CM) to the point of contact between pusher and workpiece, and the radius a of the disk which circumscribes the workpiece. When these parameters are given, the locus of centers of rotation for all possible pressure distributions can be found.

$\vec{c}$ shown are useful in considering the motion of the five-sided workpiece shown inscribed in the disk. We do not require the point of contact to be on the perimeter of the disk, as this would eliminate applicability of the results to workpieces inscribed in the disk. Indeed, for generality we do not even require the point of contact to be within the disk. Similarly, we will not require α to be such that the edge being pushed is perpendicular to vector $\vec{c}$, as it would be if the workpiece were truly a disk. The disk (with radius a), α, $\vec{c}$, and the CM are shown in Figure 3-3. A particularly simple pressure distribution $P(\vec{w})$, in which the support is concentrated at just a "tripod" of points ($\vec{w}_1$, $\vec{w}_2$, $\vec{w}_3$) is indicated, along with the COR that might result for that pressure distribution.

3.1.10. Notation

- Vectors are indicated by an arrow, for example, $\vec{v}$.
- $\vec{r}$ is the vector from the CM to the COR. r is the magnitude of that vector, that is, the distance from the CM to the COR.
- A Greek letter is used to represent both an *angle* and a *unit vector* that makes that angle with respect to the x axis (measured CCW). An arrow is used to indicate the unit vector: $\vec{\alpha} = (\cos \alpha, \sin \alpha)$.
- We indicate functional dependence with subscripts. E_r is a function of $\vec{r}$ (the COR).
- All integrals are over the area of the disk.
- Curly brackets indicate a locus of values of a quantity.

3.2. USING THE MINIMUM POWER PRINCIPLE

Suppose that the geometry of a pushing operation is specified; that is, the radius a of the disk enclosing the workpiece, the point $\vec{c}$ at which the workpiece is being pushed, and the angle α of the flat surface involved in the push. If we suppose further that a single pressure distribution is specified, then a unique COR at a single point must be the result.

Our system is *constrained* because the pusher and the workpiece are in contact, the pusher is advancing a distance dx in a given instant, and the workpiece must slide enough to accommodate the advance of the pusher. The COR could be at almost any point in the plane and still allow the workpiece to accommodate the advance of the pusher. However, some of these locations will require a greater rotation of the workpiece (about the COR) to accommodate the advance of the pusher than do others.

To solve for the COR we use the minimum power principle proven in Chapter 2. The minimum power principle states that the motion the system makes (e.g., the COR about which the workpiece actually *does* choose to rotate) will be the one for which the energy dissipated to sliding friction is minimized.

We have proven that minimum power mechanics is correct under some fairly restrictive conditions: slow (quasi-static) motion is required, and the only dissipative forces that may occur in the system are (slightly generalized) analogues of Coulomb friction. The present system qualifies.

3.3. SOLUTION FOR THE COR LOCUS

In this section we compute the energy that is dissipated due to friction when the pusher advances a distance dx, as a function of the center of rotation $\vec{r}$, and for a given pressure distribution $P(\vec{w})$. We will then minimize the energy with respect to $\vec{r}$ to find the COR about which the workpiece actually *does* choose to rotate.

It may help to imagine the disk "pinned" at the COR. This is not difficult to imagine if the COR happens to fall inside the perimeter of the disk, and one's intuition can be extended to include the case where the COR is outside the perimeter. Either way, the disk is free to rotate *only about the COR*, and the COR itself stays stationary.

Given the COR, the motion of the disk is fully determined when we apply our constraint: the edge being pushed (at $\vec{c}$) must move out of the way of the advancing pusher, but stay in contact.

3.3.1. Relation Between Motion of the Pusher and Rotation of the Workpiece

To accommodate the advance dx of the pusher, the disk will rotate an amount $d\theta$ about the center of rotation $\vec{r}$. A rotation of $d\theta$ allows an advance of the pusher dx consisting of two parts, as shown in Figure 3-4.

$$dx_1 = d\theta|\vec{c} - \vec{r}| \cos\theta = d\theta(c_y - r_y)$$

$$(3.2)$$

$$dx_2 = dx_1 \frac{\tan\theta}{\tan\alpha} = d\theta \frac{c_x - r_x}{\tan\alpha}$$

Note that dx_2 corresponds to slipping of the point of contact along the workpiece edge.

Defining the unit vector $\vec{\alpha} = (\cos\alpha, \sin\alpha)$ we can write

$$dx = dx_1 + dx_2 = \frac{d\theta}{\sin\alpha} \vec{\alpha} \cdot (\vec{c} - \vec{r})$$

$$(3.3)$$

To avoid proliferation of absolute value signs, henceforth $\vec{\alpha} \cdot (\vec{c} - \vec{r})$ will be taken to be positive.

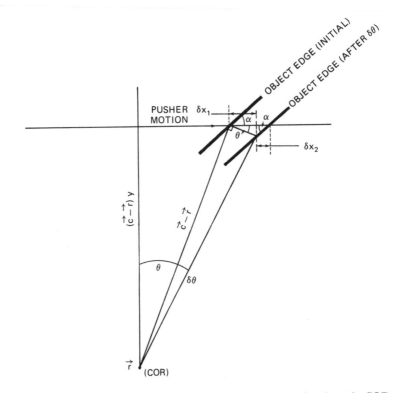

Figure 3-4 **Relation between advance of pusher (dx) and rotation about the COR ($d\theta$).** For fixed COR the pusher may advance a distance dx while the workpiece rotates an angle $d\theta$ about the COR. dx consists of two parts: movement of the workpiece edge (dx_1) and slipping of the pusher along the edge (dx_2).

Considerations of symmetry will allow application of the results to cases where $\vec{\alpha} \cdot (\vec{c} - \vec{r})$ is negative. Physically, $\vec{\alpha} \cdot (\vec{c} - \vec{r}) > 0$ corresponds to clockwise rotation of the workpiece as it is pushed.

3.3.2. Energy Lost to Friction with the Table

An area element of the disk at $\vec{w}$ supports a force $P(\vec{w})d\vec{w}$ normal to the table. The element will slide a distance

$$d\theta|\vec{w} - \vec{r}| \tag{3.4}$$

due to the rotation $d\theta$ about the center of rotation $\vec{r}$, and in the process will dissipate an amount of energy

$$dE_r = \mu_s P(\vec{w})d\vec{w} \, d\theta|\vec{w} - \vec{r}| \tag{3.5}$$

Integrating over the area of the disk, the total energy dissipated due to rotation $d\theta$ is

$$E_r = d\theta \, \mu_s \int P(\vec{w}) |\vec{w} - \vec{r}| d\vec{w} \qquad (3.6)$$

where we write E_r to remind ourselves that the energy is a function of the presumed location of the center of rotation $\vec{r}$. Substituting for $d\theta$, we have

$$E_r = \frac{dx \, \mu_s \sin \alpha}{\vec{\alpha} \cdot (\vec{c} - \vec{r})} \int P(\vec{w}) |\vec{w} - \vec{r}| d\vec{w} \qquad (3.7)$$

The system will find a location for $\vec{r}$ which minimizes E_r. At this minimum the derivatives of E_r with respect to both $\vec{r}_x$ and $\vec{r}_y$ must be zero. Evaluating the derivative of E_r with respect to $\vec{r}$ and setting it equal to zero, we find

$$\nabla E_r = dx \, \mu_s \sin \alpha \, \frac{[d_r \vec{\alpha} - \vec{v}_r \vec{\alpha} \cdot (\vec{c} - \vec{r})]}{[\vec{\alpha} \cdot (\vec{c} - \vec{r})]^2} = 0 \qquad (3.8)$$

where

$$d_r = \int P(\vec{w}) |\vec{w} - \vec{r}| d\vec{w} \qquad (3.9)$$

a scalar, can be physically interpreted as the weighted *distance* from the COR to the pressure distribution, and

$$\vec{v}_r = \int P(\vec{w}) \frac{\vec{w} - \vec{r}}{|\vec{w} - \vec{r}|} d\vec{w} \qquad (3.10)$$

a vector, can be interpreted as the weighted *direction* from the COR to the pressure distribution.

3.3.3. A Digression: Iterative Numerical Solution

Minimization of E_r can be carried out in an iterative manner to find the COR for a given pressure distribution $P(\vec{w})$. Figure 3-5 shows the locus of CORs obtained in this manner. Each point is the COR for a randomly chosen three-point pressure distribution. Only pressure distributions consisting of three points (a tripod) need be considered since according to Mason's theorem 5 [42] three points are sufficient. Weights were computed for the three points in such a way as to satisfy the constraint that the CM be at the center of the disk. (If this required any of the weights to be negative, the tripod was discarded.)

For each tripod selected as above, the center of rotation was located by this iterative method: an initial guess was made for the location of the COR $\vec{r}$, and ∇E_r evaluated at that point. The minimization technique used requires computation of $\nabla(\nabla E_r)$, the second derivative of E_r (a two-by-two matrix), which can be obtained analytically. A new guess for $\vec{r}$ is then made by adding to the old guess

$$\Delta \vec{r} = \frac{-\nabla E_r}{\nabla(\nabla E_r)} \qquad (3.11)$$

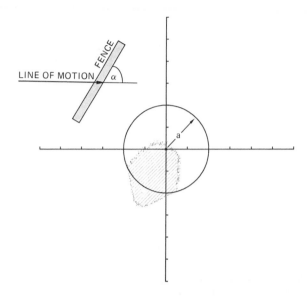

Figure 3-5   COR locus for a disk found by iterative minimization (dots). The disk shown encloses the workpiece of interest. The pusher moves horizontally along the line indicated and contacts the edge of the workpiece at the arrowhead. (In reality this point of contact would always fall within the disk bounding the workpiece, but numerical convergence is simplified for this unrealistic case.) The angle α of the edge that the pusher contacts is indicated. Dots indicate the locations of the center of rotation for 500,000 randomly chosen pressure distributions supporting the workpiece.

This method usually converged quickly if the initial guess was sufficiently close to the correct answer. By moving only one leg of the tripod at a time, and by only a small amount, the value of $\vec{r}$ found for one tripod could be used as an initial guess for the next. Figure 3-5 represents 590,000 tripods, taking 4 CPU hours on a VAX-780. Similar figures done with four points of support instead of tripods are identical, numerically validating Mason's theorem 5 [42]. Figure 3-6 is a similar run with a square workpiece replacing the disk.

3.3.4. Analytic Solution

Resuming our analytical discussion from Section 3.3.2, we set $\nabla E_r = 0$ in equation (3.8). The constant terms drop out leaving

$$r^2\vec{\alpha} = \vec{q}_r[\vec{\alpha} \cdot (\vec{c} - \vec{r})] \tag{3.12}$$

where we define the *quotient moment*, a vector, as

$$\vec{q}_r = r^2 \frac{\vec{\nu}_r}{d_r} \tag{3.13}$$

with $\vec{v}_r$ and d_r given in equations (3.9) and (3.10). $\vec{q}_r$ is a function of the COR $\vec{r}$ and the pressure distribution $P(\vec{w})$ and has units of distance. In this section we *hold the center of rotation $\vec{r}$ fixed* and analyze the quotient moment for all pressure distributions $P(\vec{w})$.

The *quotient locus* $\{\vec{q}_r\}$ is the set of $\vec{q}_r$ for all possible choices of the pressure distribution $P(\vec{w})$ consistent with the requirements listed in Section 3.1.5. It is still a function of $\vec{r}$, but the dependence on $P(\vec{w})$ has been removed. Unfortunately we have been unable to develop any physical intuition about the meaning of the quotient locus. We regard it merely as an intermediate mathematical construction, more tractable than the COR locus to which it is related.

We will always plot the quotient locus displaced by $\vec{r}$, that is, based at the COR. $\{\vec{q}_r\}$ may be plotted as a region of space, if we remember that a given $\vec{q} \in \{\vec{q}_r\}$ is a *vector* with its tail at the COR and its head anywhere in that region.

We will find the boundary of the quotient locus. The results will allow us to find the boundary of the COR locus in Section 3.3.9.

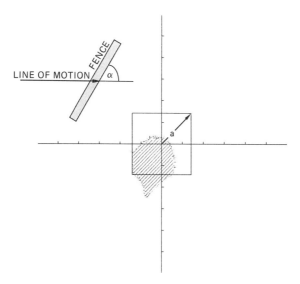

Figure 3-6   COR locus for a square found by iterative minimization (dots). It is only for a disk that we are able to find the COR locus boundary analytically. Here the CORs for a square are found numerically for 500,000 randomly chosen pressure distributions. This square can be inscribed in the disk shown in Figure 3-5, and the parameters of pushing α and $\vec{c}$ are the same. It can be seen that the COR locus (dots) found here fill an area that is entirely covered by the COR locus found for the disk (Figure 3-5). For workpieces other than disks, such as the square shown here, the COR locus for the circumscribing disk is a useful bound on the COR locus for the actual workpiece.

To simplify discussion, we take the total weight of the workpiece $Mg = 1$, that is,

$$Mg = \int P(\vec{w})d\vec{w} = 1 \qquad (3.14)$$

Since multiplying the pressure distribution P by a constant factor changes both numerator and denominator of $\vec{q}_r$ by that same factor, the assumption is harmless. Physically, the mass of the disk has no effect on the motion, so we can choose it arbitrarily.

3.3.5. Extrema of the Quotient Locus

Since $\vec{\nu}_r$ [equation (3.10)] can be interpreted as a weighted average of *unit vectors* from the COR to the pressure distribution, the greatest magnitude $\vec{\nu}_r$ can have will be 1 and will be attained when the pressure distribution is concentrated at the CM. In all other cases the direction to elements of the pressure distribution varies, and so some cancellation is inevitable. When the magnitude of $\vec{\nu}_r$ is maximal, it must be directed from the COR to the CM.

The smallest magnitude $\vec{\nu}_r$ can achieve depends on whether the COR is inside or outside the disk, that is, on whether $r > a$ or $r < a$, where a is the radius of the disk. In either case we wish to achieve the maximum amount of cancellation of direction possible. If $r > a$, this occurs when the pressure distribution consists of two points at opposite edges of the disk, providing the minimum possible agreement on direction between the two vectors, as shown in Figure 3-7.

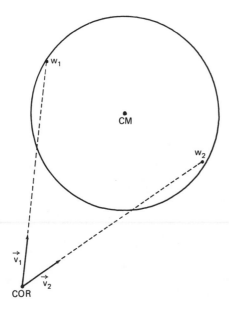

Figure 3-7 Dipod responsible for the smallest value of $\vec{\nu}_r$, for $r > a$. We study extrema of the moments $\vec{\nu}_r$ and d_r of the pressure distribution to find extrema of the "quotient moment" $\vec{q}_r = \vec{\nu}_r/d_r$. And we study extrema of the quotient moment $\vec{q}_r$ to obtain bounds on the COR to which it is related.

$\vec{\nu}_r$ is the weighted unit vector from the COR ($\vec{r}$) to the pressure distribution. It is maximized when the pressure distribution supporting the workpiece is concentrated at the CM. When $r > a$, $\vec{\nu}_r$ is minimized by the pressure distribution shown here: half the weight of the workpiece is concentrated at each of the two points of support $\vec{w}_1$ and $\vec{w}_2$, which are chosen to provide as little agreement in direction from the COR as possible.

If $r < a$, we can arrange for $\vec{v}_r$ to be zero. Indeed we can arrange for $\vec{v}_r$ to point from the COR maximally *away* from the CM by making a two-point pressure distribution as shown in Figure 3-8. (In the figure the distance from w_2 to the COR is infinitesimal.) The two vectors $\vec{w}_1$ and $\vec{w}_2$ point in opposite directions. To maintain the centroid of the pressure distribution at the CM, we find the weights of $\vec{w}_1$ and $\vec{w}_2$ are

$$P_1 = \frac{r}{r + a} \tag{3.15}$$

and

$$P_2 = \frac{a}{r + a}$$

Therefore $\vec{w}_2$ is more heavily weighted than $\vec{w}_1$, and

$$\vec{v}_r = P_1\vec{v}_1 + P_2\vec{v}_2 = (P_2 - P_1)\vec{v}_2 = \frac{a - r}{a + r}\vec{v}_2 \tag{3.16}$$

points from the COR *away* from CM.

Now consider d_r [equation (3.9)]. Clearly if the pressure distribution is concentrated at the CM, the weighted distance from the COR to the pressure distribution is just r. In fact r is the smallest value which d_r can attain. In the configuration shown in Figure 3-8,

$$d_r = P_1 \cdot (a + r) + P_2 \cdot 0 = r \tag{3.17}$$

d_r takes on its maximum value when the pressure distribution consists of two points as in Figure 3-7. That value is

$$d_r = (r^2 + a^2)^{1/2} \tag{3.18}$$

Since $\vec{q}_r$ is the quotient of $\vec{v}_r$ and d_r, extreme values of $|\vec{q}_r|$ occur when $\vec{v}_r$ is maximal and d_r minimal, and when $\vec{v}_r$ is minimal and d_r maximal. Figures 3-7

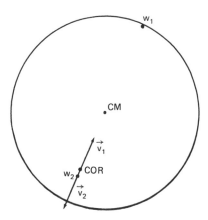

Figure 3-8   Dipod responsible for a negative value of $\vec{v}_r$, for $r < a$. If the COR is within the disk ($r < a$), it is even possible to arrange for $\vec{v}_r$ to point from $\vec{r}$ *away* from the CM, by choosing the pressure distribution to be a dipod such as this one. As w_2 is closer to the CM than w_1, it bears more than half the weight of the disk.

and 3-8 illustrate the pressure distributions that (simultaneously) minimize $\vec{\nu}_r$ and maximize d_r for $r > a$ and $r < a$, respectively.

3.3.6. Numerical Exploration of the Quotient Locus

We can find the locus of all possible quotients numerically. It is much easier to find the $\{\vec{q}_r\}$ locus (for a given value of $\vec{r}$) than it is to find the COR locus. No iteration is required; for a given tripod, the moments $\vec{\nu}_r$ and d_r can be calculated immediately. Figures 3-9 and 3-10 show typical $\{\vec{q}_r\}$ loci for $r < a$ and $r > a$, respectively. The dots are values of $\vec{q}_r$ found numerically, while the solid curve is the empirical boundary of the locus as described shortly.

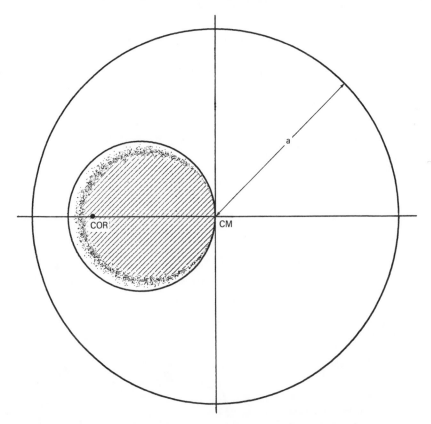

Figure 3-9   Quotient locus $\{\vec{q}_r\}$ (dots) and empirical boundary (solid), for $r < a$.
Hundreds of thousands of randomly selected pressure distributions were chosen, and for each the quotient moment was evaluated and plotted (dots). All the observed values of the quotient moment fall within the boundary (solid curve) generated by quotient moments of special pressure distributions consisting of just two points of support: dipods. In fact, the boundary turns out to be a circle, the radius of which can be determined analytically.

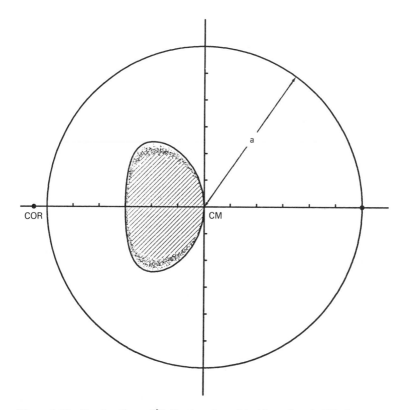

Figure 3-10   Quotient locus $\{\vec{q}_r\}$ (dots) and empirical boundary (solid), for $r > a$.
As in Figure 3-9, the quotient moments for randomly generated pressure distributions all fall within the boundary generated by quotient moments of a special group of dipods. Here $r > a$, and the bean-shaped boundary does not have a simply named shape such as the circle we found for $r < a$. However, it is still described by analytic formulas.

The dots in Figures 3-9 and 3-10 represent over 3,000,000 and 500,000 randomly chosen tripods, respectively. The solid curves which appear to bound the dots are generated by two classes of dipods, discussed shortly. On the basis of numerical studies such as shown in these figures, we believe that no value of $\vec{q}_r$ generated by a tripod or any other pressure distribution falls outside the dipod curve. Therefore, the dipod curve is the exact boundary of $\{\vec{q}_r\}$. We have not been able to prove analytically that no value of $\vec{q}_r$ falls outside the dipod curve, so the boundaries should be considered empirically justified only.

3.3.7. Boundary for $|COR| < a$

We observe that for $r < a$ the boundary of the locus is a circle. This empirical boundary can be generated by two-point pressure distributions (dipods) of the type shown in Figure 3-11, where the angle ω can vary.

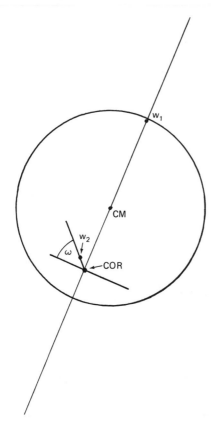

Figure 3-11 Dipods contributing to the boundary of $\{\vec{q}_r\}$**, for** $r < a$**.** When $r < a$, that is, when the COR turns out to be within the disk, these are the pressure distributions which are responsible for the boundary of the quotient locus and thus also are responsible for the boundary of the COR locus. They are simply dipods, in which one point of contact between workpiece and sliding surface is at the periphery of the disk and the other point is internal to the disk, near what turns out to be the COR. More than half the weight is supported by the internal contact, as it is nearer to the CM. It is not surprising that the workpiece rotates about a COR essentially coincident with a point supporting most of the weight of the workpiece [22]. As the internal point of support is moved in an infinitesimal circle parametrized by angle ω, the corresponding COR traces out the boundary of the COR locus inside the disk.

These dipods are a generalization of the one shown in Figure 3-8. (The distance from $\vec{r}$ to $\vec{w}_2$ is infinitesimal.) We can then calculate a parametric form for the boundary in terms of ω:

$$\vec{q}_r = \frac{r}{r + a}\left(\frac{r^2}{a}\vec{\omega} - \vec{r}\right) \tag{3.19}$$

where $\vec{\omega} = (\cos\omega, \sin\omega)$. This generates a circle of radius

$$b = \frac{ar}{r + a} \tag{3.20}$$

3.3.8. Boundary for |COR| > a

For $r > a$, the empirical boundary of the locus $\{\vec{q}_r\}$ is generated by dipods of the type shown in Figure 3-12, where ω is allowed to vary. These dipods are a generalization of the dipod shown in Figure 3-7. Again, the boundary can be calculated parametrically from ω (via intermediate terms $d^+, d^-, \gamma^+,$ and γ^-) as

$$d^\pm = (r^2 + a^2 \pm 2ar \cos \omega)^{1/2} \tag{3.21}$$

$$\sin \gamma^\pm = \frac{a \sin \omega}{d^\pm}$$

$$\cos \gamma^\pm = (1 - \sin^2 \gamma^\pm)^{1/2}$$

$$\vec{v}_r = \left(\frac{\sin \gamma^+ - \sin \gamma^-}{2}, \frac{\cos \gamma^+ + \cos \gamma^-}{2} \right)$$

$$d_r = \frac{d^+ + d^-}{2}$$

$$\vec{q}_r = r^2 \frac{\vec{v}_r}{d_r}$$

It is the boundaries of $\{\vec{q}_r\}$ that will be used (in Section 3.3.9) to determine the boundaries of the COR locus. Therefore the boundaries of the COR locus, too, can be found by considering only dipods. This is a stronger statement than Mason's theorem 5, which requires tripods. Additionally,

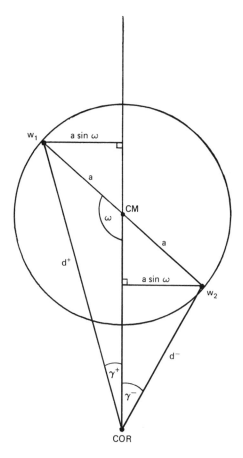

Figure 3-12 Dipods contributing to the boundary of $\{\vec{q}_r\}$, for $r > a$. When $r > a$, that is, when the COR turns out to be outside the disk, these are the pressure distributions which are responsible for the boundary of the quotient locus and thus also are responsible for the boundary of the COR locus. Again, they are simply dipods, but now in each dipod both points of contact with the sliding surface are at the periphery of the disk and so each supports half the weight of the workpiece. As the dipod system rotates around the CM (parametrized by angle ω), the corresponding COR traces out the boundary of the COR locus outside the disk.

we have found the two points constituting the dipods. However, it should be noted that the sufficiency of tripods holds for any workpiece, whereas dipods are sufficient only for a disk.

Figures 3.9 and 3.10 demonstrate that the two classes of dipods considered, and illustrated in Figures 3-11 and 3-12, generate extremal quotient moments. In other words, the locus $\{\vec{q}_r\}$ of values of $\vec{q}_r$ for all pressure distributions $P(\vec{w})$ satisfying the conditions of Section 3.1.5 fall inside the empirical boundary generated by the dipods. The boundaries themselves are, of course, part of $\{\vec{q}_r\}$, since the boundaries are generated by acceptable pressure distributions.

3.3.9. Analytic Form of the COR Locus

Having found a parametric representation of the $\{\vec{q}_r\}$ locus, we can find the COR locus. Recall the requirement for minimizing the energy lost to friction [equation (3.12)]:

$$r^2\vec{\alpha} = \vec{q}_r[\vec{\alpha} \cdot (\vec{c} - \vec{r})] \qquad (3.22)$$

The COR locus is the set of all $\vec{r}$ for which there exists a $\vec{q}_r \in \{\vec{q}_r\}$ satisfying equation (3.22).

Equation (3.22) is a vector equation. The left side obtains its direction from the edge angle $\vec{\alpha}$. The right side obtains its direction from $\vec{q}_r$, since $\vec{\alpha} \cdot (\vec{c} - \vec{r})$ is a scalar. To satisfy the vector equation $\vec{q}_r$ must have direction $\vec{\alpha}$. We can rewrite equation (3.22) in scalar form, retaining the direction constraint on $\vec{q}_r$ separately:

$$r^2 = |\vec{q}_r|[\vec{\alpha} \cdot (\vec{c} - \vec{r})] \qquad (3.23)$$

where

$$\vec{q}_r \in \{\vec{q}_r\}$$

and

$$\vec{q}_r \| \vec{\alpha}$$

We wish to find the locus of $\vec{r}$ for all distributions $P(\vec{w})$. It is best to imagine $\vec{r}$ to be an independent variable. Each value of $\vec{r}$ yields a locus $\{\vec{q}_r\}$, with one element $\vec{q}_r \in \{\vec{q}_r\}$ corresponding to each acceptable pressure distribution $P(\vec{w})$. For some values of $\vec{r}$, the value of $\vec{q}_r$ required to satisfy equation (3.23) is in $\{\vec{q}_r\}$; for other values, it is not. The former values constitute the COR locus.

It is confusing, but unavoidable, that the locus $\{\vec{q}_r\}$ shifts as we consider different locations of the center of rotation $\vec{r}$. In Figure 3-13 we have plotted several $\{\vec{q}_r\}$ loci for different values of $\vec{r}$. Note that varying the *magnitude* of $\vec{r}$ continuously changes the shape or size of the $\{\vec{q}_r\}$ loci. But

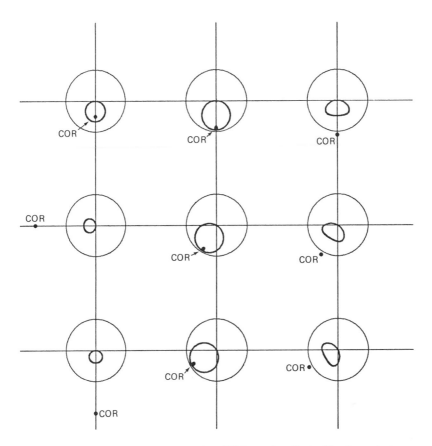

Figure 3-13   Boundaries of quotient loci $\{\vec{q}_r\}$ for various $\vec{r}$. As $\vec{r}$ is changed, the boundary of the quotient locus changes continuously. Sweeping $\vec{r}$ around the CM causes a corresponding rotation of the quotient locus boundary. Changing the distance of $\vec{r}$ from the CM changes the shape and size of the quotient locus boundary.

changing the *direction* of $\vec{r}$ only causes a corresponding rotation of the $\{\vec{q}_r\}$ locus.

The variables of equation (3.23) are shown geometrically in Figures 3-14, 3-15, and 3-16. In each figure we have plotted a value of $\vec{r}$ and the locus $\{\vec{q}_r\}$ for that $\vec{r}$. We then calculate and plot the value of $\vec{q}_r$ required to satisfy equation (3.23). In Figure 3-14, the value of $\vec{q}_r$ required does not fall in $\{\vec{q}_r\}$, so the value of $\vec{r}$ shown is not in the COR locus. In Figure 3-15, the value of $\vec{q}_r$ required *does* fall in $\{\vec{q}_r\}$, so the value of $\vec{r}$ shown is in the COR locus. In Figure 3-16, the value of $\vec{q}_r$ required to satisfy equation (3.23) happens to be on the boundary of the $\{\vec{q}_r\}$ locus. The boundary of the COR locus is generated by such cases. Interior points of the COR locus are generated when the $\vec{q}_r$ required is interior to the $\{\vec{q}_r\}$ locus, as in Figure 3-15. Since we are

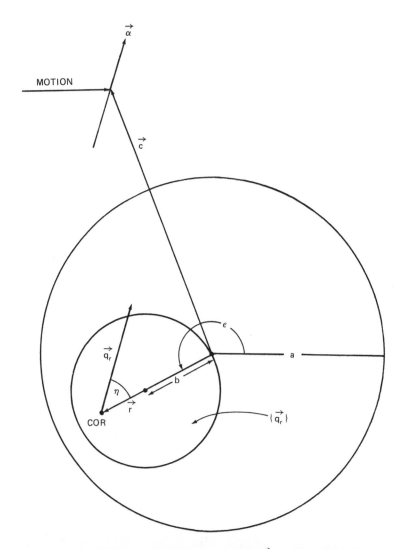

Figure 3-14   Variables of equation (3.23), for a value of $\vec{r}$ not in the COR locus. Is a proposed value of $\vec{r}$ the COR of the workpiece for some pressure distribution? First generate the quotient locus boundary for the proposed $\vec{r}$. In this case it is a circle, because $\vec{r}$ falls within the disk. Now compute the value of $\vec{q}_r$, which would be required to satisfy energy minimization [equation (3.23)]. Plot it too. If $\vec{q}_r$ falls within the quotient locus boundary (which it does not here), then $\vec{r}$ is the COR of the workpiece for some pressure distribution. $\vec{q}_r$ points to a quotient moment in the locus, so the pressure distribution which led to that quotient moment is the one which causes the COR to be at $\vec{r}$.

interested only in the boundary of the COR locus, we will consider only values of $\vec{q}_r$ which are on the boundary of the $\{\vec{q}_r\}$ locus, as shown.

3.3.10. Solution for the $|COR| < a$ Part of the COR Locus

It will be convenient to represent the COR by its polar coordinates (r, ε), and to define the relative angle η. Both angles are shown in Figure 3-16. We have

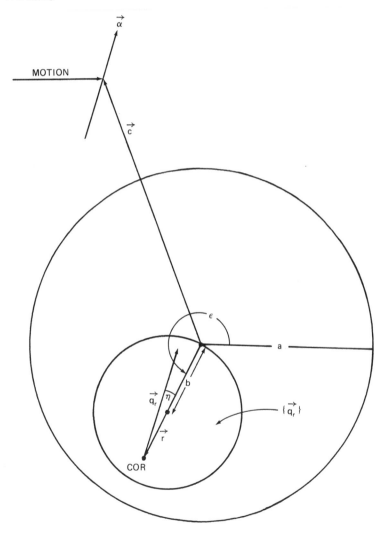

Figure 3-15   Variables of equation (3.23), for a value of $\vec{r}$ in the COR locus. Here $\vec{q}_r$ *does* fall within the quotient locus boundary, so the COR is at $\vec{r}$ for some pressure distribution.

$$\varepsilon = \pi + \alpha - \eta \tag{3.24}$$

If $r < a$, the boundary of $\{\vec{q}_r\}$ is a circle. The condition that $\vec{q}_r$ lie on the circle can be expressed

$$\left| \, |\vec{q}_r| \vec{\alpha} + (r - b)\vec{\varepsilon} \, \right| = b \tag{3.25}$$

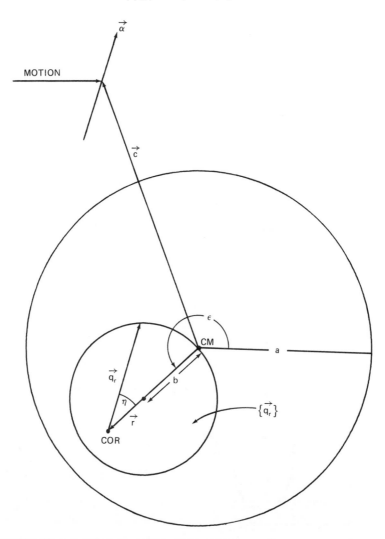

Figure 3-16   Variables of equation (3.23), for a value of $\vec{r}$ on the boundary of the COR locus. Here $\vec{q}_r$ falls on the boundary of the quotient locus, so $\vec{r}$ is on the boundary of the COR locus. We could test all values of $\vec{r}$ to see if they fall on the boundary in this way. Instead, we generate the boundary of the quotient locus (parametrized by an angle ω in the dipods) and solve for the value of $\vec{r}$ which gives rise to a $\vec{q}_r$, satisfying this figure.

where b is the radius of the circle, from equation (3.20). Equation (3.25) can be expressed in terms of the angle η as

$$[|\vec{q_r}| - (r - b)\cos\eta]^2 + [(r - b)\sin\eta]^2 = b^2 \qquad (3.26)$$

Solving this quadratic equation for $|\vec{q_r}|$ we find

$$|\vec{q_r}| = (r - b)\cos\eta \pm \{b^2 - [(r - b)\sin\eta]^2\}^{1/2} \qquad (3.27)$$

Inserting this value of $|\vec{q_r}|$ into equation (3.23) and eliminating the square root we obtain

$$\left[\frac{r^2}{\vec{\alpha}\cdot(\vec{c} - \vec{r})} - (r - b)\cos\eta\right]^2 = b^2 - [(r - b)\sin\eta]^2 \qquad (3.28)$$

Substituting b from equation (3.20) and simplifying we find

$$r^2(a + r) + (r - a)[\vec{\alpha}\cdot(\vec{c} - \vec{r})]^2 - 2r^2[\vec{\alpha}\cdot(\vec{c} - \vec{r})]\cos\eta = 0 \qquad (3.29)$$

where

$$[\vec{\alpha}\cdot(\vec{c} - \vec{r})] = \vec{\alpha}\cdot\vec{c} + r\cos\eta$$

Equation (3.29) is cubic in r and quadratic in $\cos\eta$. The solution for $\cos\eta$ is

$$\cos\eta = \frac{r[(r + a)^2 + (\vec{\alpha}\cdot\vec{c})^2]^{1/2} - a(\vec{\alpha}\cdot\vec{c})}{r(r + a)} \qquad (3.30)$$

The other quadratic root is invalid. Since η is related by equation (3.24) to the polar angle ε, equation (3.30) describes the boundary of the COR locus in the polar coordinates (r, ε) for $r < a$. A typical COR locus boundary generated using equation (3.30) is shown in Figure 3-17.

3.3.10.1. Extremal radius of the COR locus boundary for $|COR| < a$.
The minimum radius of the COR locus boundary occurs at $\varepsilon = \alpha$, which corresponds to $\eta = \pi$. From equation (3.29) we find

$$r_{min} = \frac{a(\vec{\alpha}\cdot\vec{c})}{2a + (\vec{\alpha}\cdot\vec{c})} \qquad (3.31)$$

Note that r_{min} is not the minimum distance from the CM to an *element* of the COR locus; that distance is zero. r_{min} is the minimum distance from the CM to the *boundary* of the COR locus. r_{min} is indicated in Figure 3-17.

It will also be useful to have the angles at which the COR locus boundary intersects the disk boundary. From equation (3.30) we obtain

$$\cos\eta_{r=a} = \frac{[(\vec{\alpha}\cdot\vec{c})^2 + 4a^2]^{1/2} - (\vec{\alpha}\cdot\vec{c})}{2a} \qquad (3.32)$$

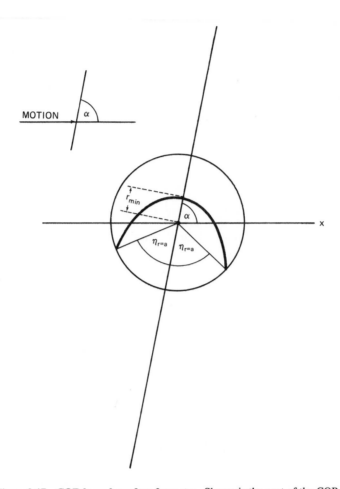

Figure 3-17 COR locus boundary for $r < a$. Shown is the part of the COR locus boundary internal to the disk. The pressure distributions which give rise to CORs on the bold boundary are dipods, with one point of support at the COR and the other on the periphery of the disk as far as possible from the COR.

r_{min} is the minimum distance from the CM to the boundary of the COR locus. Note that r_{min} is not the minimum distance from the CM to an *element* of the COR locus; that distance is zero.

3.3.10.2. Curvature of the COR locus boundary at r_{min}.   From equation (3.29) we can find the radius of curvature of the COR locus boundary at r_{min} to be

$$s = \frac{a(\vec{\alpha} \cdot \vec{c})[(\vec{\alpha} \cdot \vec{c}) + 2a]^2}{(\vec{\alpha} \cdot \vec{c})^3 + 4a(\vec{\alpha} \cdot \vec{c})^2 + 8a^2(\vec{\alpha} \cdot \vec{c}) + 4a^3} \tag{3.33}$$

3.3.11. Solution for the $|COR| > a$ Part of the COR Locus

If $r > a$, we cannot find a simple equation analogous to equation (3.25) constraining $\vec{q}_r$ to the boundary of $\{\vec{q}_r\}$. An effective approach is to parametrize the boundary of the $\{\vec{q}_r\}$ locus by the angle ω of equation (3.21) and solve for both ε and r by binary search.

For each ω the following procedure is used: we guess a value of r, in the range $a < r < r_{tip}$, where r_{tip} is an upper bound to be found in Section 3.3.11.1. Equation (3.21) is then used to calculate a value of $\vec{q}_r$. Angle η is related to the terms of equation (3.21) by

$$\eta = \arctan \frac{-\nu_x}{\nu_y} \tag{3.34}$$

and so can be computed from ω. Equation (3.23) can be written in terms of the angle η as

$$r^2 = |\vec{q}_r|(\vec{\alpha} \cdot \vec{c} + r \cos \eta) \tag{3.35}$$

which is easily tested. If it is satisfied, we have found angle η and magnitude r describing a point on the boundary of the COR locus. ε is then obtained from η using equation (3.24).

If the left-hand side of equation (3.35) is greater (resp. less) than the right-hand side, we increase (resp. decrease) the value of r guessed earlier. In this way we perform a binary search, quickly converging on a solution for r and ε.

Figure 3-18 shows the boundary of the COR locus for various $\vec{c}$ and α. The part of the boundary inside the disk was computed using equation (3.30), while the part outside the disk was found by binary search as outlined here. Calculation of each locus required about 2 CPU seconds on a VAX-780.

3.3.11.1. Tip line.

We can calculate the extremum of the COR locus analytically. For many purposes this may be all that is required. Additionally, it gives us a range within which to conduct the binary search discussed in Section 3.3.11. By symmetry, r takes on an extremal value when $\eta = 0$. In Figure 3-12 this corresponds to $\vec{\nu}_x = 0$, which in turn occurs only when $\omega = 0$ or $\omega = \pi/2$.

The extremum at $\omega = 0$ has no apparent meaning. At $\omega = \pi/2$ we find from equation (3.21)

$$\vec{q}_r = \vec{\alpha} \frac{r^3}{r^2 + a^2} \tag{3.36}$$

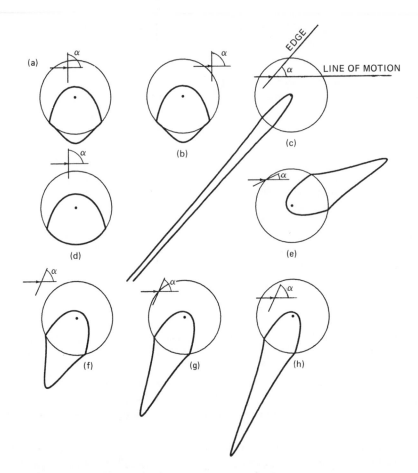

Figure 3-18 Boundaries of COR loci for various $\vec{c}$ and α. The pressure distributions which give rise to CORs on the boundary of the COR locus external to the disk are dipods, with both points of support on the periphery of the disk diametrically opposite each other. The boundary is generated as the angle ω parametrizing the dipods is varied (Figure 3-12).

In the figures the point at which the workpiece is being pushed is indicated by an arrowhead, and the angle (α) of the edge being pushed is indicated by the line the arrowhead contacts. (In several cases the arrowhead is outside the disk; this is unrealistic.)

At this value equation (3.23) yields

$$r_{tip} = \frac{a^2}{\vec{\alpha} \cdot \vec{c}} \tag{3.37}$$

This is the greatest distance $\vec{r}$ may be from the CM, and it occurs at polar angle $\varepsilon = \pi + \alpha$. In Figure 3-19 we plot r_{tip} versus contact angle α, for a given value of $\vec{c}$. As α is varied, the tip of the COR locus at distance r_{tip} from the CM traces out a straight line, the *tip line*.

The use of this graphical construction is illustrated in Figure 3-19. For a given value of α, as shown, r_{tip} is at the intersection of the tip line just described with a line through the CM at angle $\vec{\alpha}$.

An interesting case occurs when $\vec{\alpha}$ becomes perpendicular to $\vec{c}$. (Note that this does not require $\alpha = \pi/2$.) As $\vec{\alpha} \cdot \vec{c} \to 0$, we have $r_{tip} \to \infty$. The COR at infinity corresponds to pure translation perpendicular to $\vec{\alpha}$. Figure 3-18c shows a case in which $\vec{\alpha}$ is almost perpendicular to $\vec{c}$. Note that $r_{tip} \to \infty$ does not mean that pure translation is assured, only that it is possible. The COR may fall at any distance less than r_{tip}.

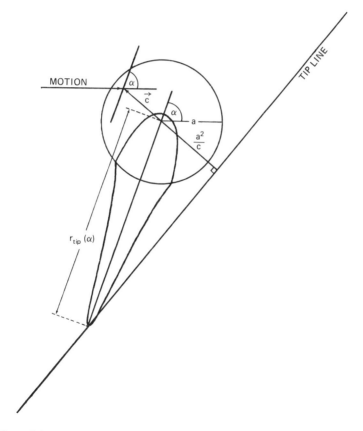

Figure 3-19 $r_{tip}(\alpha)$ **versus α and construction of the tip line.** The most useful point on the COR locus boundary seems to be the tip, as this is the COR for which rotation of the workpiece is slowest. The distance to the tip (from the CM) is given by the simple formula $r_{tip} = a^2/\vec{\alpha} \cdot \vec{c}$. As the angle of the pushed edge α is varied, the tip of the COR locus sweeps out a straight line called the tip line.

The radius of curvature of the COR locus boundary at the tip can be found analytically to be

$$s = \frac{r_{tip}}{\frac{1}{2} + (r_{tip}^4/a^4)} \tag{3.38}$$

3.3.12. Symmetries of the COR Locus

We now have the ability to quickly compute the COR locus for any $\vec{c}$ and α.

The COR locus is a function of four parameters: the disk radius a, the edge angle α (which may be the angle of the pushing fence or of the workpiece edge pushed, measured with respect to the line of motion of the pusher), and the two components of the point of contact $\vec{c}$ between pusher and pushed workpiece. However, the COR locus is really much simpler in functional dependence than the existence of four parameters would seem to imply.

The most obvious symmetry is one of total size: if both $\vec{c}$ and a are changed by a factor of γ, the COR locus will be scaled by a factor of γ as well.

Note that the COR locus has an axis of symmetry through the CM at angle $\vec{\alpha}$. The "tip" of the locus falls on this axis of symmetry, and the tip-line construction (Section 3.3.11.1 and Figure 3-19) makes use of this symmetry.

The shape of the COR locus depends *only* on the distance of the tip of the locus from the CM, $a^2/\vec{\alpha} \cdot \vec{c}$, as a multiple of the disk radius a. If COR loci for various tip distances are precomputed, we need only select the appropriate one, scale it by the disk radius a, and tilt it at the appropriate angle α.

Finally, the COR locus can depend only on the force and torque applied by the pusher. Displacing the point of contact $\vec{c}$ perpendicular to the edge angle α (i.e., along the line of action of the applied force) changes neither force nor torque, and therefore cannot change the COR locus. In Figure 3-18, the COR loci in sections a and b are identical because the point of contact $\vec{c}$ has been displaced perpendicular to the edge.

3.3.13. Summary

We have found the boundary of the COR locus for any choice of $\vec{c}$ and α. Within the disk the boundary is given by a simple formula relating r and ε, the polar coordinates of the boundary [equation (3.30)]. Outside of the disk, the polar coordinates of the boundary are found by binary search as outlined in Section 3.3.11. For most applications it is not necessary to find the entire

COR locus boundary, as simple formulas exist for several important points on the boundary. Most important of these is the tip-line construction described in Section 3.3.11.1.

Further brief discussion of the boundaries of the quotient locus (Section 3.3.7) is in order. The quotient locus is an intermediate mathematical construction whose boundaries are transformed directly into the boundaries of the COR locus. The boundaries of the quotient locus were found by making an informed guess as to the pressure distributions which give rise to the boundaries. Then this guess was tested by extensive computer simulation of random pressure distributions. These numerical results suggest that the analytic quotient locus boundaries were indeed correct: no randomly generated pressure distribution ever appeared which landed outside the analytic boundary of the quotient locus. Because of the empirical justification of the boundaries of the quotient locus, however, our derivation of the analytic boundaries of the COR locus is not rigorous. It may well be that it was this step (requiring computer testing) which prevented analytic solution for the COR locus long ago [28] [39] [58].

3.4. APPLICATION

The foregoing results can be usefully applied to the problem of aligning a workpiece by pushing it. In Figure 3-1 a misoriented rectangle is being pushed by a fence. The fence is moving in a direction perpendicular to its front edge. Evidently the rectangle will rotate CW as the fence advances [42] and will cease to rotate when the edge of the rectangle comes into contact with the front edge of the fence [12]. The problem is to find how far the fence must advance to assure that the CW motion is complete.

The geometry of this problem differs from the geometry used in previous sections. Previously a point pusher made contact with a straight workpiece edge. Here the straight edge of the pusher makes contact with a point (corner) of the workpiece. But since the coefficient of friction between the pusher and the edge of the workpiece (μ_c) is zero, we know that in either case the force exerted by the pusher on the workpiece is normal to the edge, regardless of whether the edge is that of the pusher or that of the workpiece. Since the motion of the workpiece can depend only on the force applied to it, the angle of the fence takes the place of the angle of the workpiece edge (α), and all the results derived remain unchanged.

In this section we will generalize the problem slightly, relative to the problem illustrated in Figure 3-1:

- The workpiece pushed is arbitrary, not a rectangle.
- The motion of the fence is not necessarily perpendicular to its face.

First, we circumscribe a disk of radius a about the workpiece. The disk is centered at the CM of the workpiece (Figure 3-20). Note that the contact point need not be on the perimeter of the circumscribed disk.

We know [42] that the workpiece will rotate CW and will cease to rotate when the final configuration shown in Figure 3-21 is reached.

We now ask the rate of rotation of the workpiece about the COR, with unit advance of the pusher dx. Let the angle of the CM from the direction of motion of the pusher be β. This is also the angle between the tip line and the

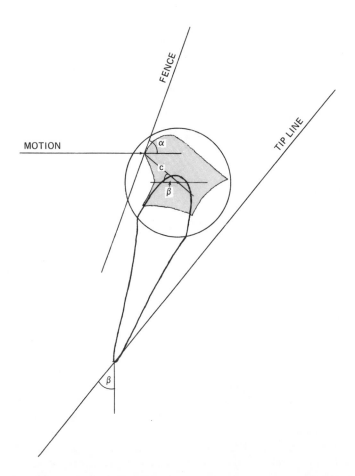

Figure 3-20 Initial configuration of workpiece and fence and resulting COR locus. The fence travels horizontally and contacts the shaded workpiece as shown. As the fence advances, the workpiece rotates clockwise at a rate which depends upon the location of the COR. The workpiece is circumscribed by a disk of radius a, since this is the only shape for which we can find exact COR locus boundaries. The ice-cream-cone shaped COR locus boundary is shown. The minimum rate of rotation occurs when the COR is at the tip of the locus.

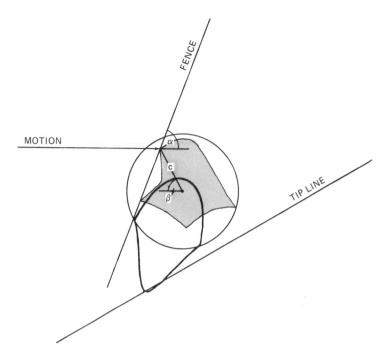

Figure 3-21   Final configuration of workpiece and fence and resulting COR locus.
Finally the workpiece has rotated into alignment with the fence. At the moment
before alignment the COR locus boundary is as shown. We want to determine the
maximum advance of the fence which could possibly be required to get from the
orientation shown in Figure 3-20 to the one shown here. So we assume that the
COR is always at the tip of the locus, which is the point at which the workpiece
rotates most slowly as the fence advances.

perpendicular to the line of motion. (Both angles are indicated in Figure 3-
20). From equation (3.3) we have

$$dx = \frac{d\beta}{\sin \alpha} \, \vec{\alpha} \cdot (\vec{c} - \vec{r}) \tag{3.39}$$

where $\vec{r}$ is the distance from the CM to the COR. The rate of rotation per
advance of the pusher, $d\beta/dx$, depends on where the COR $\vec{r}$ falls within the
COR locus. Since we wish to find the *longest* push which could possibly be
necessary to achieve a certain amount of rotation, we need to know for
which $\vec{r}$ in the COR locus $d\beta/dx$ is minimized, that is, we consider the worst
case location for $\vec{r}$. This occurs when $\vec{r}$ is at the tip of the COR locus.
Therefore, we use

$$dx = \frac{d\beta}{\sin \alpha} \, \vec{\alpha} \cdot (\vec{c} - \vec{r}_{tip}) \tag{3.40}$$

Using r_{tip} from equation (3.37), this can be integrated to yield the indefinite integral

$$x = \frac{-c \sin(\alpha + \beta)}{\sin \alpha} - \frac{a^2}{2c \sin \alpha} \log \left| \frac{1 + \sin(\alpha + \beta)}{1 - \sin(\alpha + \beta)} \right| \qquad (3.41)$$

To find the maximum pushing distance, Δx, required to cause the workpiece to rotate from its initial configuration shown in Figure 3-20 to its final configuration shown in Figure 3-21, we simply substitute the initial and final values of β into equation (3.41) and take the difference $x_{final} - x_{initial}$.

4

The COR Locus Including Contact Friction

In this chapter we present results for the locus of centers of rotation for all possible pressure distributions, in the presence of friction between the pushing and pushed workpieces. To demonstrate the use of our results, we find the distance a polygonal workpiece must be pushed by a fence to assure alignment of an edge of the workpiece with the fence. We also analyze the motion of a sliding disk as it is pushed aside by the corner of a workpiece in linear motion. Finally, we study the effectiveness of a sensorless manipulation strategy based on "herding" a disk toward a central goal by moving a pusher in a decreasing spiral about the goal.

4.1. SOLUTION FOR THE COR LOCUS INCLUDING CONTACT FRICTION

Up to now we have assumed that the coefficient of friction between the pusher and the edge of the pushed workpiece was zero, that is, $\mu_c = 0$. The pushing force was therefore normal to the edge being pushed. Since the motion of the workpiece can depend only on the force applied to it, we will designate the locus we found $\{COR\}_\alpha$ to indicate its dependence on the force angle, which is perpendicular to $\vec{\alpha}$.

We know how to generate the COR locus for a given angle of applied force. Unfortunately, when $\mu_c > 0$, it is not possible to tell what the force angle will be. We will describe angular *limits* on the force angle in Section 4.1.1, but within those limits, the force angle depends on the pressure distribution, which is not known. If we already knew that the COR would be at a certain point, however, it would then be possible to find the force angle.

Our approach to this problem is to seek CORs that are consistent with the force angle which gives rise to them. For each force angle ϕ within the angular limits, we generate $\{COR\}_\phi$. For each COR in $\{COR\}_\phi$, we find the force angle implied. If the force angle implied matches ϕ, that COR is a possible one for the workpiece. This formulation seems to threaten a great deal of computation, which in fact is not required.

We will refer to the set of consistent CORs as the *COR sketch*, to distinguish it from the elementary COR loci $\{COR\}_\phi$ produced for known force angles. Two elementary COR loci will be used in the construction of the COR sketch. In the figures, these COR loci will be left visible in outline, while the actual COR sketch—the consistent CORs—will be shown shaded.

4.1.1. Contact Friction and the Friction Cone

Let μ_c be the coefficient of friction between the pusher and the workpiece. If $\mu_c > 0$, two distinct modes of behavior of the system are possible: *sticking* and *slipping*. In Figure 3-1, sticking means that the element of the fence in contact with the corner of the workpiece remains invariant as the pusher's motion proceeds. Referring to Figure 3-2, sticking means the element of the workpiece edge which is in contact with the pushing point remains invariant as the pusher's motion proceeds. Slipping is simply the case in which either the element of the pusher or the element of the workpiece, which are in contact with each other, changes as the motion proceeds.

Define

$$\nu = \tan^{-1} \mu_c \qquad (4.1)$$

In Figure 4-1 we construct a *friction cone*, of half-angle ν, at the point of contact $\vec{c}$. The cone is centered on the edge normal, at angle $\alpha - \pi/2$ relative to horizontal. Note that the edge may be either that of a fence, where it contacts a corner of the workpiece (as in Figure 3-1), or an edge of the workpiece, where it is touched by a corner of the pusher (as in Figure 3-2). The friction cone is a well-known construction in classical mechanics.

The component of the applied pushing force tangential to the edge, $F_\parallel$, is supported by friction. Its magnitude cannot exceed $\mu_c F_\perp$, where $F_\perp$ is the component of force normal to the edge. Therefore the total applied force vector must lie within the friction cone.

If we attempt to apply a force to the workpiece edge at an angle outside of the friction cone, friction cannot support the tangential component of force. The result is slipping along the edge, and the actual applied force is directed along one extreme of the friction cone. If we apply a force within the friction cone, friction is sufficient to support the tangential component of force, and slipping will not occur: we have sticking.

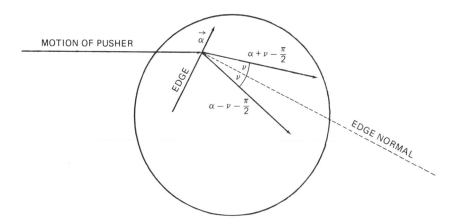

Figure 4-1 Construction of the friction cone. The force which the pusher applies to the workpiece edge must lie within the friction cone shown. If we attempt to apply a force at an angle falling outside the friction cone, friction cannot support the component of force tangential to the workpiece edge. The pusher will then slip along the workpiece edge, and the actual force applied will lie along one extreme of the friction cone. If we apply a force which lies within the friction cone, the pusher will not slip relative to the workpiece edge.

In short, slipping is only consistent with a force vector at one extreme of the friction cone, while sticking is only consistent with a force vector within the friction cone. It is not usually possible to tell if slipping or sticking will occur: often, depending on the pressure distribution, either may occur.

4.1.2. Sticking and Slipping Zones

In this section we presume that the COR is known: a single point is the COR for the workpiece. We divide the plane into three zones, called the *sticking line*, the *up-slipping zone*, and the *down-slipping zone* (Figure 4-2). The up-slipping and down-slipping zones are regions of the plane with positive areas, while the sticking line is merely a line, but all three will be collectively designated "sticking and slipping zones." The motion of the workpiece is qualitatively different for the COR falling in each of the three zones.

The sticking line is the line perpendicular to the pusher's line of motion, intersecting the point of contact between pusher and workpiece (i.e., $\vec{c}$ lies on the sticking line). Since we choose to draw the pusher's line of motion horizontally, the sticking line is vertical. The sticking line divides the down-slipping zone, on its left, from the up-slipping zone, on its right. Also shown in Figure 4-2 is the edge normal line. Above this line, the up-slipping and down-slipping designations are reversed. The area above the edge normal will be unimportant, however.

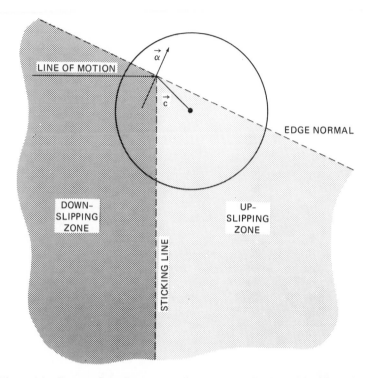

Figure 4-2   Construction of zones: up-slipping, down-slipping, and sticking line. The location of the COR has implications for slipping or sticking of the pusher with respect to the workpiece edge. If the COR lies on the sticking line shown, pusher and workpiece edge move along (horizontally) together and there is no slipping of one relative to the other. If the COR falls in the up- or down-slipping zones to either side of the sticking line, then the workpiece has a vertical component of motion and so must slip relative to the pusher (which moves horizontally).

4.1.2.1. Sticking line.   First, consider the workpiece's motion when the COR is on the sticking line. Recall that the motion of any point of the workpiece is perpendicular to the vector from the COR to that point. If the COR lies on the sticking line, the workpiece's motion at the point of contact is perpendicular to the sticking line, and is therefore parallel to the pusher's line of motion.

Since the pusher's line of motion and the workpiece's motion at the point of contact are parallel, the pusher and the workpiece, at the point of contact, travel along together. There is no need for one to slip relative to the other; the workpiece and the pusher are *sticking* at the point of contact.

4.1.2.2. Slipping zones.   Now suppose that the COR is in the down-slipping zone. The workpiece's motion at the point of contact has a downward component, relative to the pusher's line of motion. The pusher-workpiece contact must be *slipping*, with the workpiece moving down relative to the pusher.

Similarly, if the COR is in the up-slipping zone, the workpiece at the point of contact moves up relative to the pusher as the pusher advances.

4.1.3. Consistency for Slipping

If we know that the workpiece is slipping relative to the pusher (and whether up or down), then the force angle is known: it is at one extreme of the friction cone, perpendicular to $\alpha \pm \nu$.

If the COR lies in the down-slipping zone, the workpiece moves *down* as the pusher advances. Therefore the force angle must be along the upper extreme of the friction cone, at angle $\alpha + \nu - \pi/2$. Similarly, if the COR lies in the up-slipping zone, the workpiece moves *up* as the pusher advances, and the force angle must be along the lower extreme of the friction cone, at angle $\alpha - \nu - \pi/2$.

Combining these observations, we see that if slipping occurs, the COR must be either in $\{COR\}_{\alpha+\nu}$ and the down-slipping zone, or in $\{COR\}_{\alpha-\nu}$ and the up-slipping zone. These two intersection regions are called the *down-slipping locus* and the *up-slipping locus*. A very similar construction was used by Mason and Brost in Figure 5 of [45].

The down-slipping and up-slipping loci are two components of the COR sketch, because every COR in either locus is consistent with the force angle that was used to generate it. We construct the down-slipping locus of the COR sketch by intersecting the down-slipping zone (left of the sticking line) with $\{COR\}_{\alpha+\nu}$. We construct the up-slipping locus of the COR sketch by intersecting the up-slipping zone (right of the sticking line) with $\{COR\}_{\alpha-\nu}$.

In Figure 4-3, $\{COR\}_{\alpha+\nu}$ and $\{COR\}_{\alpha-\nu}$ are shown in outline. The down-slipping and up-slipping loci are the shaded areas left and right of the sticking line, respectively.

4.1.4. The Sticking Locus

The third set of consistent CORs belong to the *sticking locus*. The sticking locus, together with the up-slipping and down-slipping loci whose construction was just described, are all the CORs consistent with the force angle they presume. The three consistent loci constitute the COR sketch.

If the COR lies on the sticking line, sticking occurs. The force angle can be anywhere in the friction cone, that is, between $\alpha - \nu - \pi/2$ and $\alpha + \nu - \pi/2$. The sticking locus is therefore the intersection of the sticking line with the union, over all ϕ perpendicular to a force angle within the friction cone, of $\{COR\}_\phi$. The sticking locus is shown as a bold section of the sticking line in Figure 4-3.

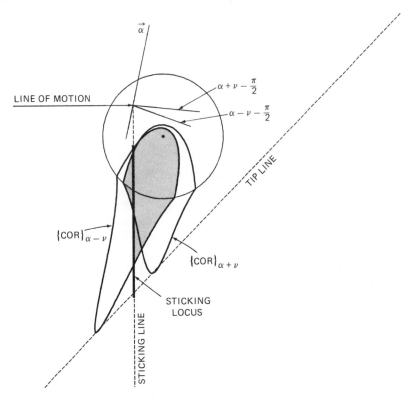

Figure 4-3 Construction of the COR sketch. When the coefficient of friction between pusher and edge of workpiece $\mu_c > 0$, the locus of possible CORs can be constructed from two of the simpler COR loci which we calculated for $\mu_c = 0$. The two $\mu_c = 0$ loci are shown in outline, while the COR "sketch" for a nonzero μ_c is shown shaded. Depending on where the COR falls in the COR sketch, slipping of the workpiece (either up or down) relative to the pusher, or sticking, may be predicted.

As discussed, the two slipping loci are $\{COR\}_{\alpha \pm \nu}$, possibly cut off by the sticking line. In calculating either slipping locus, the force angle is known: it is $\alpha \pm \nu - \pi/2$. But in calculating the sticking locus (which is just a simple line segment), the force angle is not known, except that it lies within the friction cone. To find the endpoints of the sticking locus exactly, we could form every locus $\{COR\}_\phi$, for ($\alpha - \nu < \phi < \alpha + \nu$), and intersect each locus with the sticking line. The union of these intersections is the sticking locus. This is not an efficient method.

The lower endpoint of the sticking locus is of particular interest. It is possible to approximate it by using the tip-line construction described in Section 3.3.11.1. The procedure for finding the sticking locus is to form every locus $\{COR\}_\phi$, for ($\alpha - \nu < \phi < \alpha + \nu$), and intersect each locus with the sticking line. As we vary ϕ, $\{COR\}_\phi$ varies continuously from $\{COR\}_{\alpha-\nu}$, which is outlined in Figure 4-3, to $\{COR\}_{\alpha+\nu}$, also shown outlined. The tip of the extreme loci, as well as of all intermediate loci, fall on the tip line. The tip line is shown dashed in Figure 4-3.

Were it not for the fact that each $\{COR\}_\phi$ locus drawn dips slightly below the tip line, the lower endpoint of the sticking locus would be exactly at the tip line. We will use this approximation. The small error so introduced can be bounded [54] and is usually negligible.

Using the tip line to approximate the lower endpoint of the sticking locus in this way depends on an unstated assumption: that the tip of $\{COR\}_{\alpha-\nu}$ lies to the left of the sticking line while the tip of $\{COR\}_{\alpha+\nu}$ lies to the right of the sticking line. This assumption is necessary so that the tip of some intermediate locus $\{COR\}_\phi$ will intersect the sticking line. In Section 4.1.6, we will deal methodically with this problem.

The shaded slipping loci and the bold sticking locus of Figure 4-3 contain all the possible locations of the COR.

4.1.5. Possible Configurations of an Elementary COR Locus

The down-slipping, up-slipping, and sticking *loci* play an important part in the rest of this work. It is worth describing the qualitatively different ways in which an elementary COR locus $\{COR\}_\alpha$ can intersect the three *zones* (down-slipping, up-slipping, and sticking line) to form the loci. These qualitatively different types of intersections will be called distinct *elementary configurations*. Later we will describe the qualitatively different COR sketches which can occur; they will be called distinct *sketches*. Two COR loci are used in the construction of a COR sketch, so there are more distinct sketches than distinct elementary configurations.

For a given contact point $\vec{c}$, changing α yields four distinct elementary configurations of the resulting COR loci. In Figure 4-4a, the *pure slipping* elementary configuration, the entire COR locus falls in the up-slipping zone. In Figure 4-4b, the COR intersects all three zones, but the tip of the locus falls on the same side of the sticking line as the CM. This is the *same-sided-split* elementary configuration. As α is further decreased, the tip of the COR locus crosses the sticking line, entering the *opposite-sided-split* elementary configuration, as shown in Figure 4-4c. Finally, when α decreases to the point where the edge normal at $\vec{c}$ intersects the CM, the COR locus goes to infinity [42]. The COR at infinity implies pure translation (with no rotation) of the workpiece as the pusher advances. Beyond this point the workpiece's sense of rotation switches from clockwise to counterclockwise. For our purposes in constructing a COR sketch, counterclockwise rotation is unphysical [42], and so we will class this and pure translation as one elementary configuration, the *wrapped* elementary configuration, as

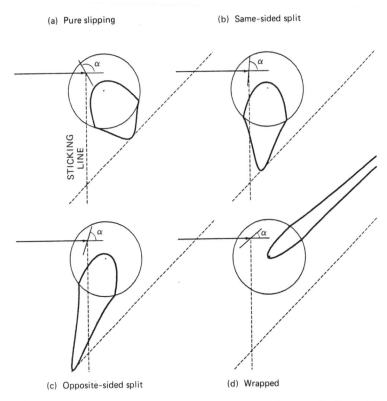

Figure 4-4 Possible elementary configurations of the COR locus. As the angle α of the pushed edge varies, the COR locus may intersect the three zones in different ways, called distinct "elementary configurations." The entire locus may fall in the down-slipping zone (part a), the locus may intersect both slipping zones and the sticking line with the tip of the locus on one side or the other (parts b and c), or the locus may "wrap" through infinity as shown in part d.

shown in Figure 4-4d. No part of a "wrapped" locus will ever contribute to the COR sketch, yet we will continue to draw its outline as shown in the figure.

The same four elementary configurations can be defined (now with increasing α) when the sticking line is to the right of the CM (Figure 4-5).

4.1.6. Possible Distinct COR Sketches

Depending on α and μ_c, each of the two elementary COR loci $\{COR\}_{\alpha \pm \nu}$ used in constructing the COR sketch may be any of the four elementary configurations described in Section 4.1.5 (pure slipping, same-sided split, opposite-sided split, or wrapped). There are nine possible distinct sketches composed of two elementary configurations, as shown in Figure 4-6. (Of the

4^2 combinations, 6 are eliminated because the tip of $\{COR\}_{\alpha+\nu}$ cannot be left of the tip of $\{COR\}_{\alpha-\nu}$. The one sketch in which both $\{COR\}_{\alpha\pm\nu}$ are "wrapped" elementary configurations is inconsistent with clockwise rotation of the workpiece.)

It is worth looking carefully at each sketch in particular to understand the construction of the sticking locus. The sticking locus is the intersection of $\{COR\}_\phi$ with the sticking line, as ϕ is swept from $\alpha + \nu$ to $\alpha - \nu$. The sweeping is always *clockwise*. In Figure 4-6g, sweeping clockwise means sweeping from the pure slipping locus, clockwise, to the wrapped locus. The intermediate loci therefore do intersect the sticking line, even though neither locus $\{COR\}_{\alpha\pm\nu}$ does. Unless this is understood, the origin of the sticking locus in Figures 4-6g and h will remain mysterious.

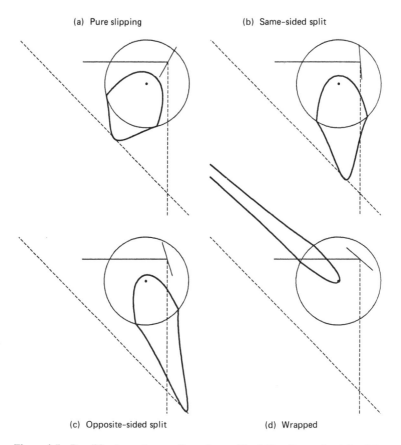

(a) Pure slipping   (b) Same-sided split

(c) Opposite-sided split   (d) Wrapped

Figure 4-5   Possible elementary configurations with sticking line to the right of the CM. The same four elementary configurations shown in Figure 4-4 can be defined when the sticking line is to the right of the CM.

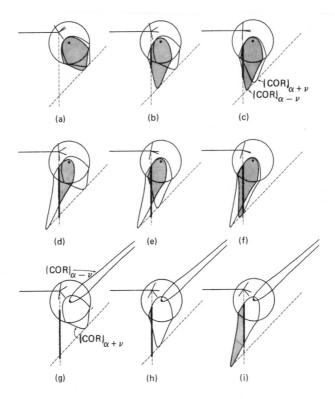

Figure 4-6 Nine distinct COR sketches with respect to the sticking line. Depending on the angle α of the pushed edge (not labeled here) and the coefficient of friction μ_c (which determines the width of the friction cones shown), the two elementary COR loci which contribute to a COR sketch may intersect the slipping and sticking zones in nine different ways.

Look closely at each distinct sketch to understand the origin of the sticking locus (the bold section of the sticking line). The sticking locus is the intersection of $\{COR\}_\phi$ with the sticking line, as ϕ is swept from $\alpha + \nu$ to $\alpha - \nu$. The sweeping is always *clockwise*. In sketch g, sweeping clockwise means sweeping from the pure slipping locus, clockwise, to the wrapped locus. The intermediate loci therefore do intersect the sticking line, even though neither locus $\{COR\}_{\alpha \pm \nu}$ does.

Several of the sketches shown in Figure 4-6 have interesting properties. In sketch a, the workpiece must slip up relative to the pusher. In sketches b and d, the workpiece must stick or slip up. In sketch g, the workpiece must stick to the pusher. In sketches h and i, the workpiece must stick or slip down. In the remaining sketches c, e, and f, either mode of slipping, or sticking, is possible, depending on the pressure distribution.

Analogous qualitative results are possible when the point of contact $\vec{c}$ is to the right of the CM. The distinct COR sketches for this case can be obtained from those shown in Figure 4-6 by reflecting about a vertical axis. (The pusher's motion should still be considered left to right, however.) The distinct sketches for counterclockwise rotation of the workpiece may be obtained by reflecting about a horizontal axis.

4.2. FROM INSTANTANEOUS MOTION TO GROSS MOTION

We have shown how to find all possible instantaneous motions of a pushed sliding workpiece, given only the parameters α, $\vec{c}$, and a. In some cases it is possible to say with certainty that a particular kind of motion, such as sticking, can or cannot occur. The set of possible CORs, as found by constructing the COR sketch, describes completely the possible instantaneous motions of the workpiece as long as those parameters remain in effect. Usually, however, the instantaneous motion that results changes the parameters (except the radius a), so that a new COR sketch must be constructed.

Often we wish to calculate not the bounds on the instantaneous direction of motion, as we did earlier, but bounds on a gross motion of the workpiece which can occur concurrently with some other gross motion of known magnitude. (For instance, we may wish to find bounds on the displacement of the pusher which occurs while the workpiece rotates 15 degrees). Our approach to dealing with gross motion follows a definite strategy, which will be illustrated in the sample problems solved in Sections 4.3, 4.4, and 4.5.

Suppose we wish to find the greatest possible change in a quantity x, while quantity β changes from $\beta_{initial}$ to β_{final}. From the geometry of the problem we find an *equation of motion* relating the instantaneous motions dx and $d\beta$. We then construct the COR sketch for each value of β. In each sketch we locate the possible COR which maximizes $dx/d\beta$. Using that COR, we integrate the equation of motion from $\beta_{initial}$ to β_{final}, yielding an upper bound for the quantity x.

Sometimes the possible COR which maximizes $dx/d\beta$ can be found analytically, or at least approximated analytically, and sometimes it must be found numerically. When an analytical solution is found, it may or may not be possible to integrate the equation of motion in closed form using that analytical solution. The examples that follow illustrate all these situations.

4.3. EXAMPLE: ALIGNING A WORKPIECE BY PUSHING WITH A FENCE

In this example, we wish to find the maximum distance a fence must advance after first contacting a workpiece, in order to assure that an edge of the pushed workpiece has rotated into contact with the fence. A typical initial configuration is shown in Figure 4-7, with the workpiece shown shaded. (Note that the fence does not advance perpendicular to its front edge.) The final configuration is shown in Figure 4-8. (In Section 3.4 we solved this problem for the case where $\mu_c = 0$.)

Also shown in Figure 4-7 is the COR sketch for the initial configuration and the angle β between the line of motion and the line from the point of contact to the CM. β is also the angle between the tip line and the sticking line. Angle β changes from 45 degrees initially in Figure 4-7 to 80 degrees in the final configuration, Figure 4-8. Note that a 1-degree rotation of the workpiece about the COR will produce a 1-degree change in β as well. We wish to find the advance x of the pusher (fence) required to change β by 35 degrees.

The workpiece's rate of rotation about the COR $d\beta$, for advance of the pusher dx, was found in equation (3.3) to be

$$dx = \frac{d\beta}{\sin \alpha} \, \vec{\alpha} \cdot (\vec{c} - \vec{r}) \qquad (4.2)$$

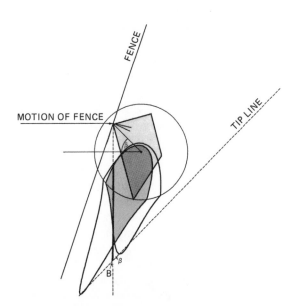

Figure 4-7 Initial orientation of the fence and pushed workpiece. As the fence advances horizontally, the four-sided workpiece rotates clockwise. The COR sketch is the shaded portion plus the bold section of the sticking line called the sticking locus. The two elementary ($\mu_c = 0$) COR loci which were used to generate the COR sketch are shown in outline. We need to find the COR responsible for slowest rotation of the workpiece. This turns out to be at the lowest point of the sticking locus (marked B), not at the tip of one of the $\mu_c = 0$ loci as in the frictionless case considered in Section 3.4.

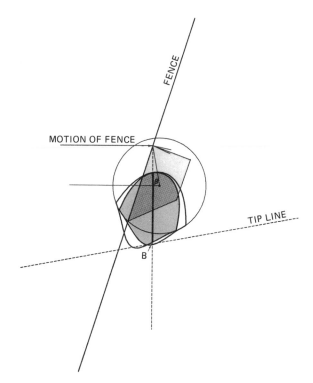

Figure 4-8   Final (aligned) orientation of the fence and pushed workpiece. Here we show the COR sketch at the moment before the conclusion of the workpiece's clockwise rotation into alignment with the fence. By this time the COR responsible for slowest rotation of the workpiece is no longer at the bottom of the sticking locus but rather at the point marked B, which is the tip of one of the elementary ($\mu_c = 0$) COR loci.

To find the maximum required pushing distance, we must find the maximum value of $\vec{\alpha} \cdot \vec{r}$ for any possible COR $\vec{r}$ in the COR sketch. This will be the *slowest* COR, the one for which the rotation of the workpiece with advance of the pusher is slowest.

Reviewing the nine distinct COR sketches in Figure 4-6, we see that the slowest COR is at the lower endpoint of the sticking locus in sketches d, e, g, and h. We will call this behavior *sticking-slowest*. It occurs when the tips of the two loci $\{COR\}_{\alpha \pm \nu}$ fall on opposite sides of the sticking line.

In sketches a, b, c, f, and i, the slowest COR is an element of one of the slipping loci $\{COR\}_{\alpha \pm \nu}$. We will call this behavior *slipping-slowest*. It occurs when the tips of the two loci $\{COR\}_{\alpha \pm \nu}$ fall on the same side of the sticking line. (For the purposes of the rule given here, the "wrapped" loci in sketches g, h, and i count as having their tip to the left of the sticking line.) In fact, the slowest COR in the slipping-slowest regime is very nearly the

COR at the tip of one of the loci $\{COR\}_{\alpha \pm \nu}$. It is only because the angle of symmetry $\alpha \pm \nu$ differs from α that the tip is *not* the slowest COR. We will use the tip of one of the loci $\{COR\}_{\alpha \pm \nu}$ as an approximation to the slowest COR. The error introduced by this approximation can be bounded [54] in terms of the radius of curvature of the tip of the COR loci, but for practical purposes it is negligible.

It is possible to have a transition from slipping-slowest behavior to sticking-slowest behavior within a pushing operation, as β increases. Such a transition occurs when the tip of one of the loci $\{COR\}_{\alpha \pm \nu}$ passes through the sticking line. In Figure 4-9, for example, it is $\{COR\}_{\alpha + \nu}$ which passes through the sticking line. We may derive the condition for the intersection:

$$a^2 + c^2 = -a^2 \tan \beta \tan (\alpha \pm \nu + \beta) \tag{4.3}$$

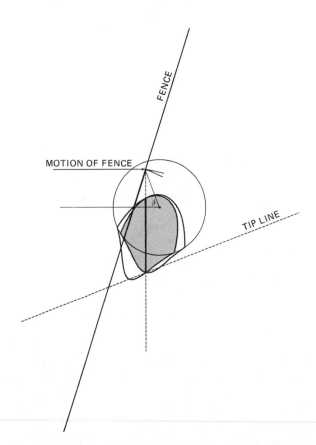

Figure 4-9   Transition from sticking-slowest to slipping-slowest behavior. This is the moment of "transition" from the COR responsible for slowest possible rotation of the workpiece being at the bottom of the sticking locus, as in Figure 4-7, to being at the tip of one of the elementary ($\mu_c = 0$) COR loci, as in Figure 4-8.

The tip of locus $\{COR\}_{\alpha \pm \nu}$ is on the same side of the sticking line as the CM when the left side of equation (4.3) is less than the right side. The value of β at which the tip crosses the sticking line may be found by solving equation (4.3) for β:

$$\tan \beta_{transition} = \frac{c^2 \tan(\alpha \pm \nu) \pm [c^4 \tan^2(\alpha \pm \nu) - 4a^2 (a^2 + c^2)]^{1/2}}{2a^2} \qquad (4.4)$$

The pushing distances required to advance β from its initial value to the transition, and from the transition to the final value, must be evaluated separately. In our example, the locus $\{COR\}_{\alpha + \nu}$ is type same-sided split initially, but changes to type opposite-sided split. Using equation (4.4) we find $\beta_{transition}$ = 69.4 degrees, as shown in Figure 4-9.

4.3.1. Slipping-Slowest Regime

If the slowest COR is at the tip of one of the loci $\{COR\}_{\alpha \pm \nu}$, we have

$$dx = \frac{d\beta}{\sin \alpha} \, \vec{\alpha} \cdot (\vec{c} - \vec{r}_{tip}) \qquad (4.5)$$

where

$$r_{tip} = \frac{a^2}{(\vec{\alpha} \pm \vec{\nu}) \cdot \vec{c}}$$

which can be integrated to yield the indefinite integral

$$x = \frac{-c \sin(\alpha \pm \nu + \beta)}{\sin \alpha} - \frac{a^2}{2c \sin \alpha} \log \left| \frac{1 + \sin(\alpha \pm \nu + \beta)}{1 - \sin(\alpha \pm \nu + \beta)} \right| \qquad (4.6)$$

Since, in the example being considered, the motion from $\beta_{transition}$ = 69.4 degrees until β_{final} = 80 degrees falls in the slipping-slowest behavior regime, we simply evaluate x at these two angles and subtract. Here the "$-$" sign in "$\alpha \pm \nu$" is used. The distance dx obtained is one component of the maximum required pushing distance to align the workpiece.

4.3.2. Sticking-Slowest Regime

In Figure 4-7 the slowest COR is the lowest point of the sticking locus, labeled B. When the COR is at point B, $(\vec{c} - \vec{r})$ may be approximated as

$$|\vec{c} - \vec{r}| = \frac{c^2 + a^2}{c \sin \beta} \qquad (4.7)$$

If the radius of curvature of the tip of the COR locus boundary were zero, this approximation would be exact. As it is not zero, the bottom of the sticking locus drops slightly below the tip line. This is a negligible effect, bounded in [54]. We will neglect it here.

Note the absence of any dependence on the friction cone angle ν. This is because when the pusher and workpiece are already sticking, further increase in μ_c has no physical effect. To find the maximum required pushing distance, it is only necessary to integrate equation (4.5) with $\vec{c} - \vec{r}$ as given here. We obtain the indefinite integral

$$x = \frac{c^2 + a^2}{2c} \log \left| \frac{1 - \cos \beta}{1 + \cos \beta} \right| \tag{4.8}$$

In our example, motion from $\beta_{initial} = 45$ degrees until $\beta_{transition} = 69.4$ degrees falls in the sticking-slowest behavior regime, so we simply evaluate x at these two angles and subtract. The distance dx obtained is the second component of the maximum required pushing distance to align the workpiece. The total required pushing distance to align the workpiece is the sum of the two partial results obtained from equations (4.6) and (4.8).

4.4. EXAMPLE: MOVING POINT PUSHING ASIDE A DISK

In this example we consider a disk being pushed not by a fence, but by a point moving in a straight line. The point may be a corner of a polygonal pusher, as long as it is only a corner of the pusher that touches the disk, and not an edge.

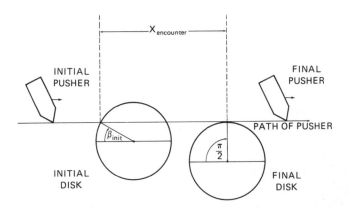

Figure 4-10   Configuration of the disk and the path of the pusher, before and after collision. A point pusher in linear motion encounters a disk. The collision is characterized by an initial value of the "collision parameter" $\beta_{initial}$. After the pusher has translated a distance $x_{encounter}$, the disk has become tangent to the path of the pusher and the two break contact, ending the collision. β_{final} is $\pi/2$. During the collision the disk rotates an angle ξ. We wish to place bounds on $x_{encounter}$ and on ξ.

In all cases the outcome of the collision is the same: the disk is pushed aside by the pusher, and contact is broken. The disk ceases to move at the instant the pusher loses contact with it (we assume slow motion), so the disk will be left tangent to the pusher's path when contact is broken. The initial and final configurations of the disk are shown in Figure 4-10. We wish to calculate the minimum and maximum length of the encounter, $x_{encounter}$, in terms of the *collision parameter*, β, as indicated in Figure 4-10. We might also wish to know the minimum and maximum angles through which the disk may rotate during the collision.

4.4.1. Length of the Encounter

In Figure 4-11, the variables of interest are x, which parametrizes the advance of the pusher along its path, and β, which completely characterizes the collision. β will vary from $\beta_{initial}$, its value at first contact, to $\beta_{final} = \pi/2$ when contact is broken. $x_{encounter}$ is the corresponding change in x, as β changes from $\beta_{initial}$ to $\pi/2$.

If the instantaneous COR is known, the direction of motion of the CM of the disk is known: it makes an angle θ with the horizontal, as shown in Figure 4-11. If the CM of the disk moves a distance Δl along its line of motion, we can find the resulting values of $\Delta\beta$ and dx, and thereby relate $\Delta\beta$ and dx to each other.

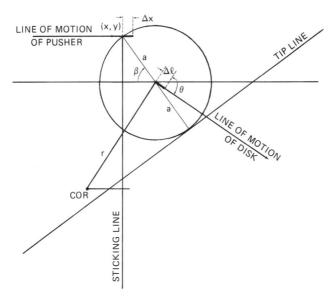

Figure 4-11   Finding equation of motion (4.9). If the COR were known, we could find relations among: (a) the motion of the CM of the disk Δl, (b) the change in the collision parameter $d\beta$, and (c) the advance of the pusher dx.

The pusher advances a distance

$$dx = \Delta l \cos \theta + \Delta l \sin \theta \tan \beta \qquad (4.9)$$

due to Δl. At all times β can be found from

$$a \sin \beta = y + \Delta l \sin \theta \qquad (4.10)$$

where (x, y) are the coordinates of the point of contact.

Substituting Δl from equation (4.9), and evaluating the change in $\sin \beta$ due to Δl, we find

$$a\Delta(\sin \beta) = \frac{dx \sin \theta}{\cos \theta + \sin \theta \tan \beta} \qquad (4.11)$$

For infinitesimal motions $\Delta\beta$ and dx becomes $d\beta$ and dx. Using $d(\sin \beta) = \cos \beta \, d\beta$, we find an equation of motion

$$dx = a \, d\beta \left(\sin \beta + \frac{\cos \beta}{\tan \theta} \right) \qquad (4.12)$$

Since it will turn out that $\tan \theta > 0$, the largest and smallest values of $dx/d\beta$ will result when θ assumes its smallest and largest values, respectively.

Now we construct the COR sketch, shown in Figure 4-12. Since the edge normal at $\vec{c}$ passes through the CM, the extremes of the friction cone pass to either side of the CM, for any $\mu_c > 0$. $\{COR\}_{\alpha-\nu}$ is a "wrapped" locus (as described in Section 4.1.5), so the COR sketch must be that of Figure 4-6 sketch g, h, or i. In any case there must be a sticking locus, there cannot be an up-slipping locus, and there may or may not be a down-slipping locus. In Figure 4-12 we have shown a down-slipping locus.

In Figure 4-12, and in general when the COR sketch is any one of distinct types g, h, or i, the smallest and largest values of θ (Figure 4-11) occur when the COR is at the lower or upper endpoints, respectively, of the sticking locus. For sketches g and h the lower endpoint of the sticking locus is well approximated by the intersection of the sticking line with the tip line, and we will use this approximation (neglecting the small effect of the curvature of the tip, though this could be included). For the lower endpoint of the sticking locus in sketch i, and for the top of the sticking locus in all three sketches, numerical methods would have to be used. We will not find these numerical results here.

4.4.1.1. Greatest length of encounter.
As in Section 4.3.2, we will neglect the slight dip of the sticking locus below the tip line, which results from the nonzero radius of curvature of the tip of the COR locus boundary.

We will also assume that the COR sketch is of type g or h, not i, so that the lower endpoint of the sticking locus can be approximated by the intersection of the sticking line with the tip line. This assumption will be addressed in Section 4.4.1.2.

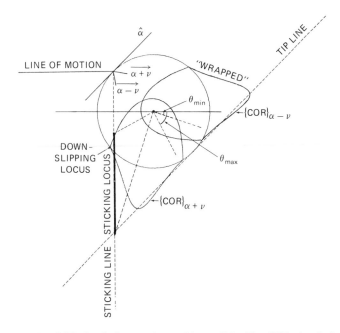

Figure 4-12   COR sketch for a point pushing a disk. The COR sketch for the collision between pusher and disk. The angle of the edge being pushed, α, is the tangent to the disk at the point of contact $\vec{c}$. Therefore one of the two elementary COR loci which compose the COR sketch is "wrapped" (Figure 4-4). The COR sketch consists of only a down-slipping locus (left of the sticking line) and a sticking locus. This is reasonable: it would be surprising if the disk should slip up relative to the pusher.

If the COR is at the intersection of the sticking line with the tip line, we find from Figure 4-13

$$\tan \theta = \frac{x_{COR}}{y_{COR}} \tag{4.13}$$

$$x_{COR} = -a \cos \beta$$

and

$$y_{COR} = a \frac{\sin^2 \beta - 2}{\sin \beta}$$

where y_{COR} is found from the construction of Figure 4-13. Using $c = a$, equation (4.13) can be simplified to

$$\tan \theta = \frac{\cos \beta \sin \beta}{1 + \cos^2 \beta} \tag{4.14}$$

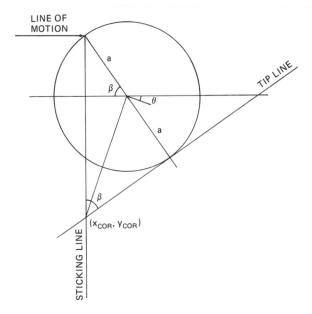

Figure 4-13 Finding the smallest θ, equation (4.13). The length of the encounter between pusher and disk is greatest if the COR is at such a location that θ is minimal. In most cases the bottom of the sticking locus is the location of the COR which minimizes θ. Using the tip line construction we can find the minimum value of θ as shown here.

Using this value of $\tan \theta$ in the equation of motion (4.12) results in

$$dx = a \, d\beta(\sin \beta) + \frac{1 + \cos^2 \beta}{\sin \beta} \tag{4.15}$$

which, integrated, yields the indefinite integral

$$x_{encounter} = a \left(\ln \frac{1 - \cos \beta}{1 + \cos \beta} \right) \tag{4.16}$$

The maximum value of $x_{encounter}$ can be obtained by evaluating equation (4.16) at $\beta_{initial}$ and $\beta_{final} = \pi/2$ and subtracting. The value at $\pi/2$ is zero.

4.4.1.2. Condition for sketch type i. The foregoing derivation of maximum $x_{encounter}$ assumed that the lower endpoint of the sticking locus is at the tip line. This is not true when the COR sketch is of type i, in Figure 4-6.

The COR sketch is of type i when the tip of $\{COR\}_{\alpha+\nu}$ is left of the sticking line. Simplifying equation (4.3) for $a = c$ and $\alpha + \beta = \pi/2$, we find the condition for sketch i to be

$$\tan \beta > 2 \tan \nu = 2\mu_c \tag{4.17}$$

This means that the COR sketch will always become type i as $\beta \to \pi/2$, unless $\mu_c = \infty$. ($\mu_c = \infty$ can occur, for example, in pushing a gear, if a tooth is engaged by the pusher.) In every case of pushing aside a disk, sketch i is entered eventually.

By using the tip line as the lower endpoint of the sticking locus, despite the fact that this is a poor approximation in sketch i, we find too low a value for the minimum θ. Our calculated maximum for $x_{encounter}$ [equation (4.16)] is unnecessarily high. We could in principle refine the upper bound by finding the lower endpoint of the sticking locus more accurately by numerical methods.

As mentioned, we are also neglecting the slight dip of the sticking locus below the tip line (in sketches g and h), which causes us to underestimate the maximum possible value of $x_{encounter}$. Here too we could refine $x_{encounter}$ by numerical methods.

Neglect of sketch i, and neglect of the dip due to tip curvature, cause errors of opposite sign in calculating the maximum $x_{encounter}$. The latter is a smaller error. Neither error will be addressed here.

4.4.1.3. Least length of encounter.   The minimum possible value of $x_{encounter}$ occurs when the COR is at the top of the sticking locus. We do not have an analytical method of finding or approximating the upper endpoint of the sticking locus, as we have for the lower endpoint. The lower endpoint is similarly hard to analyze if the COR sketch is of type i in Figure 4-6. In these cases it is necessary to find the endpoints numerically for all β in the range of interest, calculate θ for each β, and then integrate equation (4.12) numerically to find $x_{encounter}$.

4.4.2. Rotation of the Pushed Disk During Encounter

4.4.2.1. Maximum rotation.   In Section 4.4.1, both the largest and smallest possible values of $x_{encounter}$ resulted from CORs on the sticking line. If the COR remains on the sticking line, the pusher does not slip relative to the surface of the disk, and so evaluation of the rotation of the disk during the encounter, $\xi_{encounter}$, is trivial. We have

$$\xi_{encounter} = a \, \frac{\pi}{2 - \beta_{initial}} \tag{4.18}$$

Since only up-slipping of the pusher is possible, equation (4.18) is an exact upper bound for $\xi_{encounter}$; any slipping will only serve to reduce the rotation of the disk.

Maximal slipping is obtained if $\mu_c = 0$. The pushing force is directed through the CM of the disk, so the disk can only translate and not rotate [42]. So if $\mu_c = 0$, we have $\xi_{encounter} = 0$ as both maximum and minimum rotation.

4.4.2.2. Minimum rotation.   We found in Section 4.4.1 that extreme values of $dx/d\beta$ occur when θ takes on extreme values. Having constructed the COR sketch, we found that the extreme values of θ for possible CORs are assumed when the COR falls at the top or bottom of the locus. In this section we will not be able to find a single geometric variable, analogous to θ, whose extremes correspond to extremes of the rate of rotation.

Rotation of the disk will be measured by the angle ξ, measured at the COR, as shown in Figure 4-14. We can relate $\Delta\xi$ to advance of the pusher dx:

$$dx = l \sin \xi \, d\xi \qquad (4.19)$$

Combining equation (4.19) with equation (4.12), which relates $\Delta\beta$ to dx, we find

$$dx = \frac{a \, d\beta[\sin \beta + (\cos \beta/\tan \theta)]}{l \sin \xi} \, d\beta \qquad (4.20)$$

We can eliminate θ and $l \sin \xi$ in favor of the coordinates of the COR,

$$\tan \theta = \frac{x_{COR}}{y_{COR}} \qquad (4.21)$$

$$l \sin \xi = a \sin \beta - y$$

yielding

$$\frac{d\xi}{d\beta} = \frac{a(y_{COR} \cos \beta + x_{COR} \sin \beta)}{x_{COR}(a \sin \beta - y_{COR})} \qquad (4.22)$$

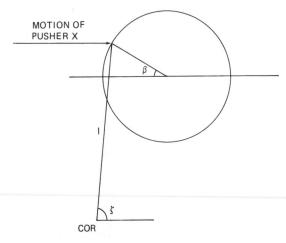

Figure 4-14 Finding equation of motion (4.19). If the location of the COR is known, the rotation of the disk ξ can be related to the advance of the pusher dx.

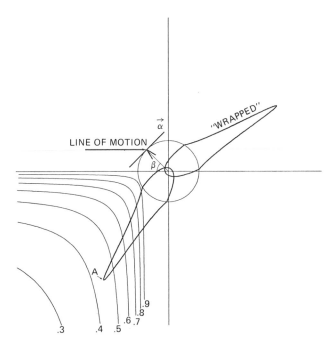

Figure 4-15 Contours of constant $d\xi/d\beta$ and the COR sketch. To find the minimum possible rotation of the disk ξ during its encounter with the pusher we seek that location of the COR which minimizes ξ for unit increase in the collision parameter β, that is, which minimizes $d\xi/d\beta$. Plotted are contours of constant $d\xi/d\beta$. We must find numerically the point in the COR locus which intersects the least contour. For the COR locus plotted, the least contour intersected is about .46, and the COR which intersects it is, once again, very near the tip of the COR locus.

This has no simple geometric interpretation. Contours of constant $d\xi/d\beta$ are plotted in Figure 4-15, for $\beta = 45$ degrees. Minimum rotation occurs at minimum $d\xi/d\beta$. The COR sketch for $\beta = 45$ degrees is superimposed on Figure 4-15. The possible value of the COR which is responsible for minimum rate of rotation is the point of the COR locus which intersects the slowest-valued contour line, indicated in the figure as point A (in this case very close to the tip). Having obtained numerically the minimum possible value of $d\xi/d\beta$, as a function of β, we can numerically find the indefinite integral:

$$\xi_{min} = \int \left(\frac{d\xi}{d\beta}\right)_{min} (\beta) \, d\beta \qquad (4.23)$$

Minimum rotation in a given collision can then be evaluated by subtracting $\xi_{min}(\beta_{initial})$ from $\xi_{min}(\beta_{final} = \pi/2)$.

4.5. EXAMPLE: SPIRAL LOCALIZATION OF A DISK

In this example we analyze an unusual robot motion by which the position of a disk (a washer, say), free to slide on a tabletop, can be localized without sensing. If the disk is known initially to be located in some bounded area of radius b_1, we begin by moving a point-like pusher in a circle of radius b_1. Then we reduce the pusher's radius of turning by an amount Δb with each revolution, so that the pusher's motion describes a spiral. Eventually the spiral will intersect the disk (of radius a), bumping it. We wish the disk to be bumped toward the center of the spiral, so that it will be bumped again on the pusher's next revolution. If the spiral is shrinking too fast, however, the disk may be bumped *out* of the spiral instead of toward its center, and so the disk will be lost and not localized.

We wish to find the maximum shrinkage parameter Δb consistent with guaranteeing that the disk is bumped into the spiral, and not out. (Δb will be a function of the present spiral radius.) We also wish to find the number of revolutions that will be required to localize the disk to some radius b, with $a < b < b_1$, and the limiting value of b, called b_∞, below which it will not be possible to guarantee localization, regardless of number of revolutions.

4.5.1. Analysis

Suppose the pushing point has just made contact with the disk. Since the previous revolution had radius only Δb greater than the current revolution, the pusher must contact the disk at a distance at most Δb from the edge of the disk, as shown in Figure 4-16. We will consider only the worst case, where the distance of the pusher from the edge is the full Δb.

We know that if $\Delta b < a$, the disk will move downward [42]. This is not sufficient to assure that the disk will be pushed into the spiral (rather than out of the spiral), because the pushing point will also move down, as it continues along its path (Figure 4-16). To guarantee that the disk will be pushed into the spiral, we must make sure that it moves down *faster* than does the pushing point.

Note that we will continue to draw the pusher's motion as horizontal, even though the pusher must turn as it follows the spiral. This is done to maintain the convention for COR sketches used in previous sections. At every moment we simply choose to view the system from such an angle that the pusher's motion is horizontal.

One way of comparing rates of moving down is by considering the increase or decrease in the angle β, called the *collision parameter,* in Figure 4-16. If, as the pusher's motion along its spiral progresses, β increases, then the disk is being pushed *into* the spiral; localization is succeeding. When β reaches $\pi/2$, the pusher grazes the disk and leaves it behind. The disk is

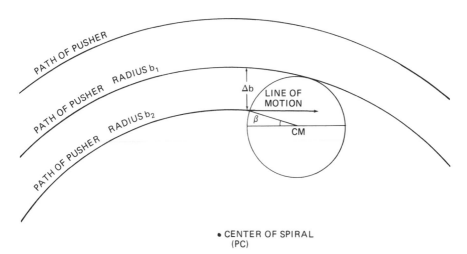

Figure 4-16 Geometry at the moment of the second collision of pusher and disk.
A point pusher describes a decreasing spiral about a region of radius b_1 within
which a disk of radius a is known to be. As the spiral decreases in radius, the disk
is pushed toward the center of the spiral. We wish to find the fastest-shrinking
spiral which will guarantee that the disk is always pushed in toward the center and
never out of the spiral. It turns out there is a limiting radius of the spiral below
which further confinement of the disk cannot be guaranteed, no matter how slowly
the spiral decreases in radius.

In this figure the disk was first struck by the pusher when it was at radius b_1 and
was pushed toward the interior of the spiral. The disk was left tangent to the path
of the pusher and is about to be struck again by the pusher, which is now at radius
b_2. $\Delta b = b_1 - b_2$ is the shrinkage rate of the spiral. Notice the collision parameter
β which results.

then left tangent to the spiral. If, as the pusher's motion progresses, β
decreases, the disk is being pushed *out* of the spiral; localization is failing.

4.5.2. Critical Case: Pusher Chasing the Disk Around a Circular Path

In the critical case the angle β does not change with advance of the
pusher. The pusher "chases" the disk around the spiral, neither pushing it
in nor out. In this section we will take the spiral to be a circle (i.e., $\Delta b = 0$),
to simplify analysis. The critical case, shown in Figure 4-17, is highly unstable. The pusher's motion is shown as an arc of a circle, labeled path of
pusher. (Underlined names refer to elements of Figure 4-17). The center of
that circle is labeled PC (for pusher center). Point PC is directly below the
point of contact, in keeping with our convention of drawing the pusher's line
of motion horizontal.

To maintain the critical case, the path followed by the CM of the disk
(labeled critical path of CM) must be as shown in the figure: an arc of a

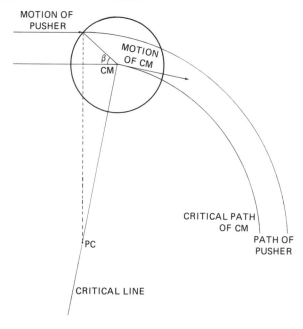

Figure 4-17 Critical case: pusher "chasing" disk around a circular path. If the shrinkage of the spiral Δb is too great, the disk can be pushed *out* of the spiral. To find the critical value of Δb below which the disk is guaranteed to be pushed *into* the spiral, we consider the marginal case where it is possible for the pusher to "chase" the disk, with the collision parameter β neither increasing (meaning that the disk is going toward the interior of the spiral) nor decreasing (meaning that the disk is going toward the exterior of the spiral).

circle, concentric with the arc path of pusher. Instantaneously, the direction of motion of the CM must be along the line labeled motion of CM, tangent to the critical path of CM. The critical line, drawn through PC and CM, is by construction perpendicular to motion of CM. The COR of the disk must fall on the critical line, in order that the instantaneous motion along the line motion of CM be tangent to the critical path of CM.

We have just seen that the COR of the disk must fall on critical line for the instantaneous motion of the CM to be consistent with the CM following the critical path of CM. If the COR falls to the left of the critical line, the CM diverges from the critical path of CM by moving *inside* the arc. Therefore β will increase with advance of the pusher, and localization is succeeding. If the COR falls to the right of the critical line, the CM diverges from the critical path of CM by moving *outside* the arc. Therefore β will decrease with advance of the pusher, and localization is failing. The critical line divides the plane into two zones: if the COR falls in the left zone, the disk is pushed into the pusher circle, while if the COR falls in the right zone, the disk is pushed out of the pusher circle.

We wish to find a condition on the radius of the pusher circle which guarantees that the disk will always be pushed *into* the circle. We will construct the COR sketch, and then find positions for PC such that all possible CORs are to the left of the critical line.

In Figure 4-18 we have constructed the COR sketch with collision parameter β. Since the edge normal at $\vec{c}$ passes through the CM, the extremes of the friction cone pass to either side of the CM, for any $\mu_c > 0$. $\{COR\}_{\alpha-\nu}$ is a "wrapped" locus (Section 4.1.5), and the COR sketch must be that of Figure 4-6g, h, or i. In any case, there must be a sticking locus, there

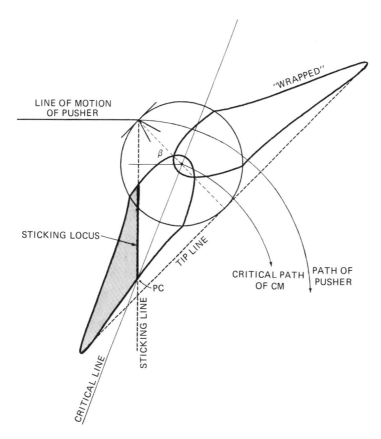

Figure 4-18   COR sketch for critical case and solution for location of PC. We wish to find a condition on the radius of the pusher circle which guarantees that the disk will always be pushed *into* the circle. We will construct the COR sketch and then find positions for PC such that all possible CORs are to the left of the critical line.

To make sure that the whole COR locus falls to the left of critical line, we need only place the center of the pusher motion (PC) *below* the lower endpoint of the sticking locus.

cannot be an up-slipping locus, and there may or may not be a down-slipping locus. In Figure 4-18 we have a down-slipping locus.

To make sure that the whole COR locus falls to the left of the critical line, we need only place the center of the pusher motion (PC) *below* the lower endpoint of the sticking locus. (Point PC is required to have the same x coordinate as the point of contact, in keeping with our convention of drawing the pusher's line of motion horizontal.)

4.5.3. Critical Radius Versus Collision Parameter

For every value of β (the collision parameter), we compute the distance from the pusher's line of motion to the lower endpoint of the sticking locus. This defines a critical radius $r^*(\beta)$. For each collision parameter β, $r^*(\beta)$ is the radius of the tightest circle that the pusher can describe with the guarantee that the disk will be pushed into the circle, or at worst be "chased" around the circle indefinitely, but not be pushed out of the circle. In Figure 4-19, $1/r^*(\beta)$ is plotted as a function of collision parameter β for each of several values of μ_c. (The discontinuity in slope results from the discontinuity in slope of the COR locus boundary at $r = a$.)

The inverse of the function $r^*(\beta)$ will be denoted $\beta^*(r)$, representing the smallest value of β for which a pusher motion of radius r still results in guaranteed localization. In terms of the pusher's distance from the top of the disk, d (Figure 4-18), we can use the relationship

$$a(1 - \sin \beta) = d \qquad (4.24)$$

to define the *critical distance from grazing* $d^*(r)$ as a function of r. $d^*(r)$ is the largest distance of the pusher from the top of the disk for which a pusher motion of radius r still results in guaranteed localization.

4.5.4. Limiting Radius for Localization

If there is a limiting radius b_∞ of the spiral motion below which localization cannot be guaranteed, then as the spiral approaches radius b_∞ the motion must become circular. $\Delta b \to 0$ as b_∞ is approached, so collisions become grazing collisions, and we have the distance from grazing $d \to 0$. (In terms of the collision parameter β, we have $\beta \to \pi/2$). The COR sketch for $\beta = \pi/2$ is shown in Figure 4-20. If the disk is not to be bumped out of the spiral, we must have $b_\infty = r^*(\beta = \pi/2)$. b_∞ is indicated in the figure and can be shown analytically to be

$$b_\infty = a(\mu_c + 1) \qquad \text{for } \mu_c \le 1 \qquad (4.25)$$

$$b_\infty = 2a \qquad \text{for } \mu_c \ge 1$$

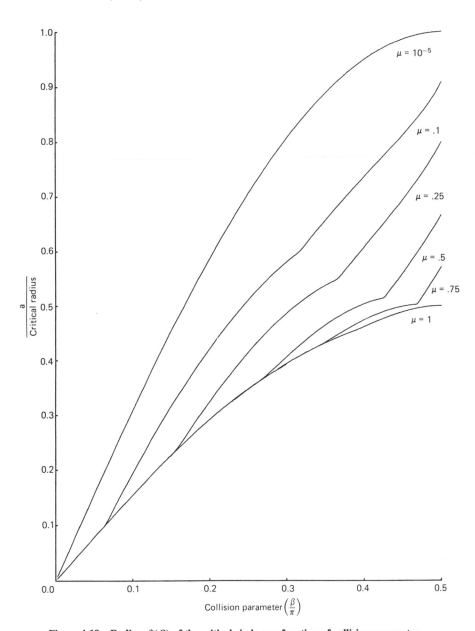

**Figure 4-19   Radius $r^*(\beta)$ of the critical circle as a function of collision parameter
β.** For every collision parameter β (here plotted as β/π), there is a tightest radius
r^* which the pusher can describe still maintaining the guarantee that the disk can
be chased or pushed inward, but never be pushed outward. For a variety of
coefficients of friction μ_c we plot here the *inverse* of that tightest (critical) radius,
a/r^*.

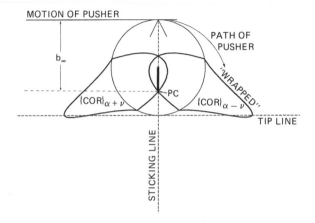

Figure 4-20   COR sketch at the limiting radius, showing b_∞. There is a limiting radius of the spiral b_∞, below which we cannot guarantee that the disk will be pushed inward, no matter how slowly the spiral is decreasing in radius, that is, no matter how small Δb. As the spiral approaches this radius it must more and more accurately approximate a circle, since it cannot go below radius b_∞. Thus the collision parameter β becomes $\pi/2$ as radius b_∞ is approached, and all collisions become grazing collisions. Drawing the COR sketch for a grazing collision we find that $b_\infty = a(\mu_c + 1)$, a general kinematic limitation on the success this herding strategy can achieve.

Only at $\mu_c = 0$ can a disk be localized completely, that is, localized to within a circle the same radius as the disk. Otherwise the tightest circle within which the disk can be localized is given by equation (4.25).

4.5.5. Computing the Fastest Guaranteed Spiral

Let b_n be the radius of the nth revolution of the pusher, so that we have initially radius b_1, and b_∞ is the limiting radius as $n \to \infty$. (In specifying but a single radius for each revolution of the spiral, we will not truly specify the spiral completely, but this will be sufficient to characterize the number of revolutions required to achieve a desired degree of localization.)

To excellent approximation we can define the fastest spiral recursively by

$$b_n = b_{n-1} - d^*(b_n) \tag{4.26}$$

The difference between the radii of consecutive turns of the spiral $n - 1$ and n is therefore $\Delta b = d^*(b_n)$. Equation (4.26) thus enforces the condition that on the nth revolution, the value of d is exactly the critical value for circular pushing motion of radius b_n. At worst, the disk is pushed neither in nor out of the spiral. A slightly slower spiral would guarantee that the disk cannot be chased in this way for long, but is pushed into the spiral. However, the difference between our spiral and the "slightly slower" one is so slight that it is not worth dealing with here [54].

Figure 4-21 shows the fractional deviation of spiral radius b_n above b_∞ versus number of turns n, on logarithmic and on linear scales. We start (arbitrarily) with $b_1 = 100a$. The spiral radius was computed numerically for $\mu_c = .25$, using the results for $\beta^*(r)$ shown in Figure 4-19 and equation (4.26).

Figure 4-21 shows that when the spiral radius is large compared to the disk radius a (which is taken to be 1 in the figure), we can reduce the radius of the spiral by almost a with each revolution. As the limiting radius is

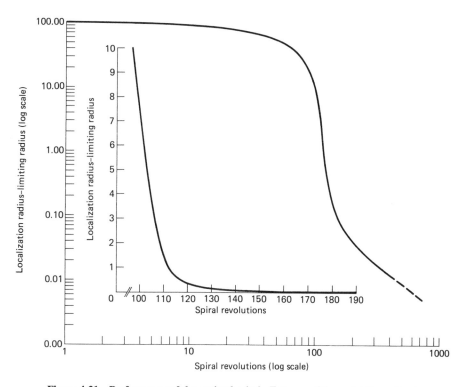

Figure 4-21   Performance of the optimal spiral. For $\mu_c = .25$ we plot the optimal spiral. This is the fastest-decreasing spiral which still guarantees that the disk is pushed into the spiral and cannot be pushed out, that is, localizes the disk as quickly as possible. We found that the spiral cannot decrease below a radius b_∞ while maintaining the guarantee, so that value has been subtracted from the vertical (spiral radius) axis, leaving only the difference between the spiral radius and its limiting value. In the linear plot, we can see that the radius of the optimal spiral decreases swiftly by almost the disk's radius a with each revolution until quite close to the limiting radius. It is then more instructive to look at the log-log plot to see how the spiral radius approaches the limiting radius.

approached, the spiral reduces its radius more and more slowly, approaching the limiting radius b_∞ as about $n^{-1.6}$, where n is the number of revolutions.

Figure 4-21 demonstrates the best performance that the "herding" strategy can achieve.

4.6. CONCLUSION

We have solved for the possible instantaneous motions of a sliding workpiece as it is pushed, in the presence of unknown frictional forces between workpiece and table, and between workpiece and pusher. We have characterized the qualitatively different kinds of sliding motion which are possible and found the conditions under which each can occur. Using these results it is possible to find bounds for *gross* motions of a pushed workpiece as well. This is done by integrating the possible instantaneous motions.

As an example, we have found the maximum distance a polygonal sliding workpiece must be pushed by a fence to guarantee that a side of the workpiece has aligned itself with the fence. Using the useful *tip-line construction* described here, approximate results are obtained both for the alignment problem and several others. Strict upper bounds for the maximum required pushing distance are found by using slightly more sophisticated methods, but the difference between the upper bounds and the approximate results are so slight that the effort seems hardly justified.

In a second example, we have taken the pushed workpiece to be a disk, and the pusher to be a point, or the corner of a polygon, moving in a straight line. We have found the maximum distance that the pusher and the disk may be in contact, before the disk is "pushed aside" by the moving workpiece. Bounds on the rotation of the disk during its interaction with the pusher are also found.

Finally, we have analyzed an unusual robot maneuver in which a disk known to be within a certain circular area can be "localized" to a much smaller circular area by a pusher which, perhaps under robot control, describes a decreasing spiral around the disk. Thus the disk can be located by a robot without sensors. We found the ultimate limiting radius below which the disk cannot be localized further, no matter how slowly the spiral decreases in radius. We also found (to within tight bounds) the "optimal spiral": the spiral that localizes the disk with the fewest number of revolutions, while guaranteeing that the disk is not lost from the spiral.

5

Planning Manipulation Strategies

In this chapter a *configuration map* is defined and computed, mapping all configurations of a workpiece before an elementary manipulative operation to all possible outcomes. Configuration maps provide a basis for planning operation sequences, which may be considered to be parts-feeder designs or sensorless manipulation strategies for robots. Sequential operations are represented as products of configuration maps for the individual operations. Efficient methods for searching the space of all operations sequences are described.

As an example we consider a class of passive parts-feeders based on a conveyor belt. Workpieces arrive on the belt in random initial orientations. By interacting with a series of stationary fences angled across the belt, the workpieces are aligned into a unique final orientation independent of their initial orientation. The planning problem is to create (given the shape of a workpiece) a sequence of fences which will align that workpiece. Using configuration maps, we transform the planning problem into a purely symbolic one. The space of all fence sequences is searched to find a successful feeder design. Designs for several workpieces are found.

5.1. APPLICATION TO INTERACTION WITH A FENCE

Consider a fence in linear motion which strikes and pushes a workpiece. For a given initial orientation of the workpiece, a particular point will be first struck by the fence. Whether a clockwise or a counterclockwise mode of

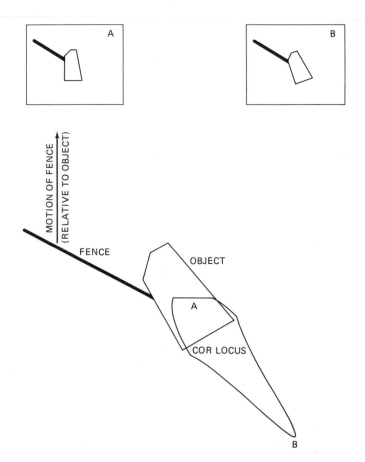

Figure 5-1 Extremal outcomes of interaction of workpiece with endpoint of fence. A snapshot of a workpiece as it leaves the endpoint of a fence. The resulting COR locus is shown. Depending upon where the COR falls within the COR locus, any final orientation between that shown in inset A and that shown in inset B may occur.

rotation then occurs can be determined from the COR locus, but if this is the only information required, it is found more simply by using the rules derived by Mason [44].

As the fence advances, the workpiece rotates, and it may also slip along the fence. The rates of rotation and slipping as the fence advances are bounded by the COR locus. The bounds allow calculation of the maximum distance the fence must advance to assure that an edge of the workpiece has rotated into alignment, and the distance the workpiece has slipped along the fence during alignment.

Finally, the workpiece leaves the end of the fence (Figure 5-1). Two points of the COR locus (shaded) in the figure are of particular interest. If the COR is anywhere along the line labeled A, the workpiece will rotate without slipping relative to the fence. Rotation without slipping may persist until a face of the workpiece is aligned with the motion of the belt, as shown in inset A. Point B gives another extreme of the possible motions of the workpiece, in which the workpiece slips relative to the fence as fast as possible for each increment of rotation. Maximal slipping may persist until the workpiece loses contact with the fence, as shown in inset B. The extreme orientations shown in the two insets define the range of possible outcomes as the workpiece interacts with the end of the fence. Line A and point B have simple analytic forms. The motion of the workpiece specified by point B can be integrated to find the extreme possible final orientation of the workpiece shown in inset B.

5.2. CONFIGURATION MAPS

The physics of an operation (for instance, a collision between a fence and a workpiece) may be encapsulated in a *configuration map*. A configuration map is a function of two copies of configuration space [36] (C-space × C-space), taking on logical values. A workpiece lying on a tabletop has a three-dimensional configuration space: it has two positional degrees of freedom and one rotational degree of freedom. The configuration map is therefore a function of six dimensions. Often, however, not all the degrees of freedom are of equal interest. For many purposes (e.g., planning a conveyor belt–based parts-aligner), all we care about is the orientation of the workpiece before and after its collision with a fence (or some other operation.) So while the configuration map representation is quite general, we will use here only a two-dimensional projection of it.

Figure 1-8 shows the configuration map M_{-60} for the workpiece and operation (interaction with a -60-degree fence) shown. We will consider the workpiece to be on a moving surface traveling downward, but it could equally well be on a stationary surface with the fence moving upward. The horizontal axis of the map gives the initial orientation θ_i of the workpiece, before it contacts the fence. (Outlines of the workpiece illustrate the orientations at several points along the axis.) The vertical axis gives the final orientation θ_f of the workpiece after it has collided with the fence, rolled along the fence until a stable edge comes into contact with the fence, and finally slid down the fence and off the end. A point $M_{-60}(\theta_i, \theta_f)$ is shown shaded if it is nonzero (logical 1), where nonzero values indicate it is possible for a workpiece with initial orientation θ_i to emerge with orientation θ_f from its interaction with the fence.

For any initial configuration of the workpiece, the configuration map gives the final configuration. Note that in the case shown, for a single initial configuration, there is a range of final configurations. This does not reflect a deficit in our physical understanding of the operation. The "one-to-many" mapping occurs because we are given only the outline of a workpiece and do not know the distribution of the weight of the workpiece upon the surface it slides on. The behavior of the workpiece depends on the distribution of weight, which in turn depends on the generally unknown details of the surfaces in contact. Using our results from Chapter 4, the set of final orientations for all distributions of weight is calculated. The horizontal bands in this configuration map are associated with discrete alignments of the polygonal faces of the workpiece.

The utility of the configuration map representation lies in the ease with which configuration maps for sequential operations can be calculated. In Figure 5-2, a workpiece is being carried along a belt, and will interact first with a −60-degree fence, and then with a +60-degree fence. A configuration map can be created which maps the workpiece's initial configuration before colliding with the first fence, into its final configurations after leaving the second fence. That configuration map is simply the matrix product of the configuration maps for the two individual interactions. In the figure the two maps to be multiplied and their product are shown. The product M_{+60-60} is

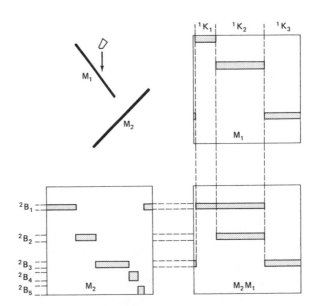

Figure 5-2 Product of two configuration maps. A workpiece is being carried along a belt, and will interact first with a −60-degree fence, and then with a +60-degree fence. A configuration map can be created which maps the workpiece's initial configuration before colliding with the first fence, into its final configurations after leaving the second fence. That configuration map is simply the matrix product of the configuration maps for the two individual interactions.

defined as

$$M_{+60-60}(\theta_i, \theta_f) = M_{+60}M_{-60} = \bigvee_\alpha \{M_{+60}(\alpha, \theta_f) \wedge M_{-60}(\theta_i, \alpha)\} \qquad (5.1)$$

where $\wedge$ is logical intersection (and) and $\bigvee$ is logical union (or).

5.2.1. Symbolic Encoding

To take advantage of the "bands" evident in the configuration map, we construct N subintervals B_j of the θ_f axis, each bounding one of the bands. For each band B_j a kernel K_j of the θ_i axis is defined as

$$K_j = \bigcup_{\alpha \in B_j} \{\theta_i | M(\theta_i, \alpha) > 0\} \qquad (5.2)$$

which is the set of initial configurations that lead to a final configuration in the band B_j. A new "rectangularized" map

$$M' = \bigcup_j K_j \times B_j \qquad (5.3)$$

is nonzero wherever M is nonzero, and perhaps at other locations as well. When M is made up entirely of rectangular bands, as in Figure 1-8, we have $M' = M$.

Now consider a product of two maps $M_2 M_1$ (operation M_1 followed by operation M_2). Using superscript 1 or 2 to indicate correspondence with one of the maps, we can express the product $M_2 M_1$ in terms of the bands of M_2 and the kernels of M_1:

$$M_2 M_1 = \bigcup_j {}^2B_j \times \{\bigcup_{k \in {}^{21}C_j} {}^1K_k\} \qquad (5.4)$$

where

$${}^{21}C_j = \{k | {}^2K_j \cap {}^1B_k \neq \varnothing\}$$

The code sets ${}^{21}C_j$ contain all the information about the product. In the example shown in the figure, with the bands B_j as labeled, we have $C_1 = \{1, 2\}$, $C_2 = \{2\}$, $C_3 = \{3\}$, $C_4 = C_5 = \varnothing$. (In figures, this code set would be written $1, 2 \rightarrow 1, 2 \rightarrow 2, 3 \rightarrow 3$.) Further products can be computed using the code sets only, for example, the code sets ${}^{321}C_j$ for the product $M_3 M_2 M_1$ are

$${}^{321}C_j = \bigcup_{k \in {}^{32}C_j} {}^{21}c_k \qquad (5.5)$$

5.3. PLANNING OPERATIONS SEQUENCES

5.3.1. The Search Tree

The space of all operations sequences may be represented as a tree. Arcs correspond to operations, for example, collisions with fences of various angles in our example. The root is labeled with the set of all initial configurations. Each node of the tree is labeled with the set of possible configurations of a workpiece after execution of the operations on the path from the root to that node. In Figure 5-3, part of a tree for operations which are collisions with fences of various angles is shown. The possible configurations of a workpiece at a given node are obtained by multiplying the configuration maps for the operations on the path from the root to that node. The product maps for the six nodes shown along the left edge of Figure 5-3 are shown in Figure 5-4. Traversing the tree in order to search it is facilitated by the ease with which products of multiple configuration maps can be computed using the code sets C_j. In Figure 5-3, each arc is labeled with a fence angle α as well as the code sets for that fence angle. The sets of

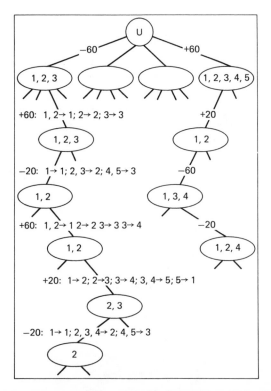

Figure 5-3 Tree for searching for an effective operations sequence. The space of all operations sequences may be represented as a tree. Arcs correspond to operations, for example, collisions with fences of various angles in our example. The root is labeled with the set of all initial configurations. Each node of the tree is labeled with the set of possible configurations of a workpiece after execution of the operations on the path from the root to that node. A goal node is one in which the set of possible configurations has been reduced to one.

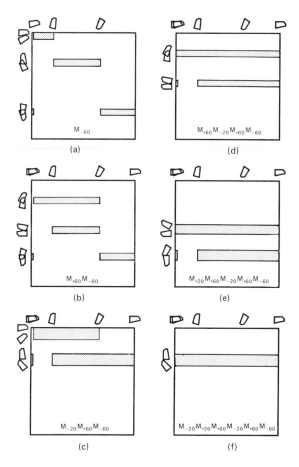

Figure 5-4 Configuration map products of successive fences. Part a is the configuration map for the interaction of the workpiece with a −60-degree fence. Part b is the configuration map for its interaction with a −60-degree fence followed by a +60-degree fence, considered as a unit. Part f is the configuration map for its interaction with all six fences considered as a unit. The map has only one final-orientation band, showing that the initial randomly oriented workpiece has been reduced to (nearly) one final unique orientation.

possible configurations which label a node are indicated as a subset of the indices j of the bands $^{\alpha}B_j$ for the fence angle α of the arc above it.

A goal node is one in which the set of possible configurations has been reduced to one, or to a sufficiently narrow range. In Figure 5-3, therefore, a goal node is labeled with just one band index i, such as is the depth 6 node on the left.

5.3.2. Pruning the Tree

Searching the tree exhaustively (to any reasonable depth, e.g., six) is essentially impossible because the branching factor at each node is so high. Two techniques may be used to make the search practical.

Among the many arcs (fences) leaving a node, only a few distinct code sets will be observed. Suppose the arcs for fence angles +50 through +60 share code sets 1, 2 → 1, 2 → 2, 3 → 3. It turns out (a result of the physics) that the final-orientation bands of a given fence are entirely contained in the corresponding final-orientation bands of a less steep fence. If a solution exists using the +55-degree arc from a given node, it must also exist using the +60-degree arc from that node. Given several arcs *having common code sets*, it is always safe to follow only the arc for the steepest fence.

The pruning step just described keeps the branching factor to a manageable level (typically 6–12). It is worth noting that for any particular incoming arc to a node, the collection of distinctly coded outgoing arcs can be precomputed.

Second, branches of the tree can be pruned while the tree is being searched. For each fence angle α, a list is kept of all node values computed after traversing an arc labeled α. If a node value is computed which is a superset of a previous value on the list, at a greater depth, the branch may be pruned. As an example, consider the depth four node labeled "1, 2, 4" in Figure 5-3. It follows an arc for a fence angle of −20 degrees. A previously visited node of depth three labeled "1, 2" also follows a −20-degree arc. "1, 2, 4" is a superset of "1, 2." If a solution exists in n steps from "1, 2, 4," the same solution must also exist from "1, 2." The total depth of the solution will be less starting from "1, 2," so it is pointless to follow the "1, 2, 4" branch further.

5.4. EXAMPLE: AUTOMATED DESIGN OF A PARTS-FEEDER

Figure 1-9 shows a top view of a system of fences suspended across a conveyor belt. The configuration map for the workpiece shown with the first (−60 degree) fence was given in Figure 1-8. The configuration map for the first two fences, considered as a unit, was given in Figure 5-2. The configuration map for the entire system of six fences is shown in Figure 5-4f. Figures 5-4a–e are the partial products as labeled.

The configuration map for the system (Figure 5-4f) has but one final-orientation band. Therefore the system of fences (Figure 1-9) is a parts-feeder: workpieces in any initial orientation emerge from their interaction with the system of fences in but one range of final orientations. To reduce

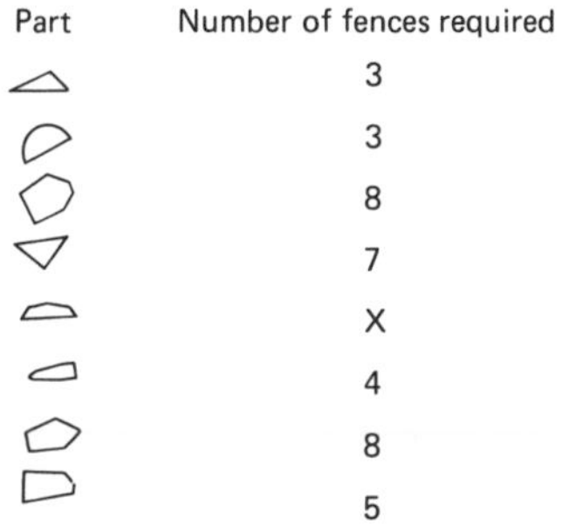

Part	Number of fences required
△	3
◌	3
◌	8
▽	7
◌	X
◌	4
◌	8
◌	5

Figure 5-5   Number of fences required in a parts-feeder for workpieces of various shapes. For each workpiece shape shown, the smallest number of fences for which a guaranteed parts-feeder design was found is indicated. For one shape, no parts-feeder design was found in a search to a depth of 20 fences.

the range of final orientations to a single final orientation, the workpieces can be "used" (e.g., picked up by a robot) before they leave the final fence.

Some workpieces have two (or more) indistinguishable orientations. A rectangle, for instance, has two. For such workpieces the configuration map of a parts-feeder has two (or more) final-orientation bands. A goal node of the search tree would be labeled with two (or more) band indices j.

5.4.1. Some Solutions

Figure 5-5 shows several workpieces and the lowest number of fences for which a parts-feeder was found. In one case, no feeder design was found in a search to a depth of 20. Planning a feeder requires only a few seconds of computation.

6

Experiments

In this chapter several experiments are described that test the adherence of real physical systems to the analytic bounds derived in earlier chapters.

6.1. MOTIVATION

The analytic work in previous sections made several assumptions about the environment in which sliding operations take place. The most important of these are

- The sliding motions are slow (quasi-static), so that frictional forces dominate inertial forces.
- The frictional force between the workpiece and the surface it slides on obeys Coulomb's law. The coefficient of friction must be velocity independent and uniform over space.

Our objective in this chapter is to demonstrate some systems in which these assumptions are valid and to show that the analytic bounds found are indeed obeyed. Some systems studied experimentally violate the bounds, and it is useful to speculate why. Finally, it is interesting to see whether the bounds derived are much broader than the range of behaviors of real systems.

In considering practical applications it is necessary to know how well the analytic bounds are obeyed. If, for instance, the real behavior of systems always falls well inside the bounds, then plans based on the bounds will result in safe, guaranteed behavior of real workpieces. On the other hand, if real systems approach and even sometimes exceed the bounds, plans should not be made which depend too critically on the bounds.

6.2. APPARATUS

All experiments were performed by a PUMA 250 robot. The robot's gripper was removed and replaced by a piece of 1-inch extruded aluminum "L" ($\frac{1}{8}$ inch thick), which was used as a fence in the experiments. The fence was glued to a $\frac{1}{2}$-inch-thick aluminum disk which was bolted to the robot wrist.

The robot's working surface was covered by a sheet of .020-inch stainless steel. On top of this other sliding surfaces could be used. The sliding surfaces used were (1) stainless steel, (2) 180 grade emery cloth, and (3) $\frac{1}{16}$-inch neoprene friction surface belting (rubberized fabric), which is a typical conveyor belt material.

At 106 cm above the sliding surface a CCD camera with a 75mm lens pointed straight down. Its field of view was approximately 11.5 cm by 15 cm. The work space was illuminated by oblique lighting from the fluorescent ceiling lights, as well as by oblique incandescent light. Little light came from angles higher than 45 degrees so that specular reflections from the steel surface into the camera could be minimized. For the emery cloth and rubberized fabric surfaces, which are dark, reflections were not a problem.

The output of the CCD camera was analyzed by a Machine Intelligence Corporation VS-100 vision system. The VS-100 divides its 256-by-256 pixel view into light and dark regions at an adjustable threshold level. It then identifies the largest light-colored blob and computes its center of mass and orientation.

The sliding workpieces were brass disks of radius 1.25 inches and brass squares of side 2 inches (radius 1.414 inches). All workpieces were $\frac{1}{4}$ inch thick. The results of previous chapters require that the height of the point of contact above the tabletop be minimal, so as not to shift the center of friction away from the CM. The motion of the fence was arranged so that the fence edge touched the sliding workpiece no more than $\frac{1}{8}$ inch from the tabletop. The upper surface of each workpiece was painted black, and a 1-inch by 2-inch white paper ellipse was glued to it. An ellipse was used so that the vision system could measure the orientation of the disk, as well as its center. An ellipse was preferred over a rectangle to minimize any possible aliasing problems within the vision system.

It is necessary to inform the vision system of the size ratio of x coordinate pixels to y coordinate pixels, which may be camera dependent. This was done by causing the system to analyze a white disk of 1.8-inch diameter on the sliding surface. The disk was placed in 10 locations, and the results averaged. The vision system x and y scales were then adjusted appropriately. The vision system was calibrated in inches, to maintain consistency with the machined workpieces which have fractional inch dimensions.

By rigidly attaching a workpiece to the robot, and repeatedly moving the robot and measuring the ellipse with the vision system, a combined measure of robot/vision system accuracy could be obtained. Angular measurements had a standard deviation of .09 degrees (1.6 mrad). Linear measurements had a standard deviation of .005 inches (.13 mm).

Both the vision system and the robot have serial lines through which a user can give instructions and receive results. Both of these lines were attached to a computer so that experiments could be performed and data recorded automatically.

6.3. STICKING LOCUS

In this experiment the front edge of the fence and the perimeter of a brass disk were coated with strips of 320 grade emery cloth to increase the coefficient of friction μ_c to about .50 (see Section 6.4). The fence angle was set at $\alpha = 70$ degrees (20 degrees from dead-on) (see Figure 6-1). In principle at this angle and coefficient of friction, the disk should not slip relative to the fence, but should always roll. (Rolling is the same as sticking.) The COR should not fall below the tip line, which intersects the sticking line at $y = -5.51a$, where a is the disk radius of 1.25 inches (Figure 6.1).

The speed of the fence could be varied, as could the sliding surface material. For each set of conditions, several hundred short pushes of about 1-cm length were performed, and the CORs recorded.

In Figure 6-2 the sliding surface is a sheet of 180 grade emery cloth. The fence was moved at 2 cm/sec. A dot shows the location of each COR observed. The vertical line is the sticking line, while the diagonal line is the tip line. If no slipping occurs, all the dots will fall on the sticking line. But even if some slipping does occur, all the dots should fall above the tip line. In this case very little slipping is observed, and all the CORs do fall above the tip line, in agreement with the analytic bounds. A histogram of the y component of the CORs is shown horizontally.

In Figure 6-3 the speed of the fence is increased to 5 cm/sec. Since this speed is still small compared to characteristic speeds for quasi-static sliding (Section 6.8), we expect (and observe) no dramatic change in the distribution of CORs.

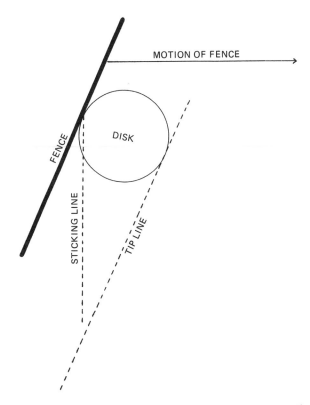

Figure 6-1 Experiment for studying the sticking locus. In this experiment the fence angle $\alpha = 70$ degrees. For the coefficients of friction μ_c used, only sticking should occur. (For a disk, sticking is equivalent to rolling.) So all observed CORs should fall along the sticking line, above the tip line.

Rubberized fabric is a particularly interesting sliding surface, because it is often used as a conveyor belt material. Results of 250 pushes on this material (at 2 cm/sec) are shown in Figure 6-4. Five CORs fall outside the tip line.

On a steel surface very poor agreement with calculated bounds is observed. In Figure 6-5 the disk is sliding on a clean steel surface, and in Figure 6-6 on a steel surface lubricated with silicone spray (Krylon 1325). In both cases it was subjectively easy to detect non-Coulomb frictional forces when sliding the disk by hand. Strangely, often the disk would resist rotating much more than it resisted translating. At higher speeds (> 20 cm/sec) the disk would sometimes trap air beneath it, which greatly reduced fric-

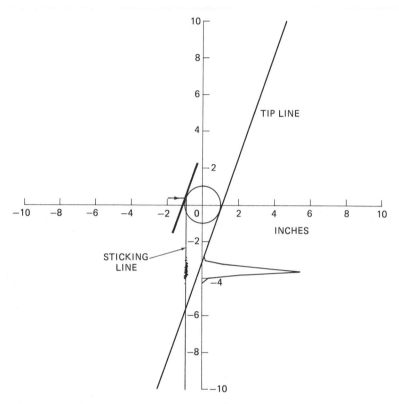

Figure 6-2 CORs for a brass disk sliding on emery cloth, while sticking to the fence. With the fence velocity chosen to be 2 cm/sec, the CORs observed in numerous repetitions of the experiment are indicated as dots. All fall on the sticking line above the tip line, as predicted. The disk was brass, sliding on emery cloth.

tion. The trapped air would sometimes abruptly disappear. In cases where either the sliding surface was more rough (as with emery cloth or rubberized fabric) or where the bottom surface of the workpiece was not flat, fewer such problems could be expected.

6.4. COEFFICIENTS OF FRICTION

The coefficient of dynamic friction between a brass disk and several sliding surfaces was determined by tipping the surface until the disk, once nudged into motion, could continue to slide down the surface. The coefficient of friction $\mu_s = \tan^{-1}(\theta)$, where θ is the angle at which sliding first occurs. Uncertainty in identifying the point at which sliding occurs creates an error of $\pm.02$ in μ_s.

μ_s	Materials
.48	Brass on 600 grade emery cloth
.37	Brass on medium grade sandpaper
.47	Brass on 320 grade emery cloth
.49	Brass on 180 grade emery cloth
.54	320 grade emery cloth on 320 grade emery cloth
.21	Brass on stainless steel
.45	Brass on rubberized fabric

For some experiments it is necessary to know the coefficient of friction between the aluminum fence and the brass edge of the workpiece μ_c. This was found by tipping the fence at increasing angles while pushing a brass square. When the fence angle exceeds $\tan^{-1}(\mu_c)$, the square should begin to slip down the fence as it is pushed along the sliding surface.

In Figures 6-7 and 6-8 the amount of slip down the fence is plotted as a function of fence angle, for pushing speeds of 2 cm/sec and 10 cm/sec, respectively. The sliding surface was 180 grade emery cloth in both cases.

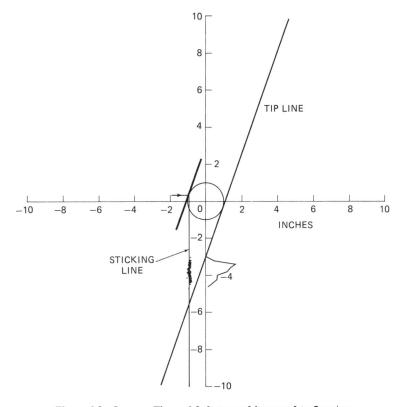

Figure 6-3   Same as Figure 6-2, but speed increased to 5 cm/sec.

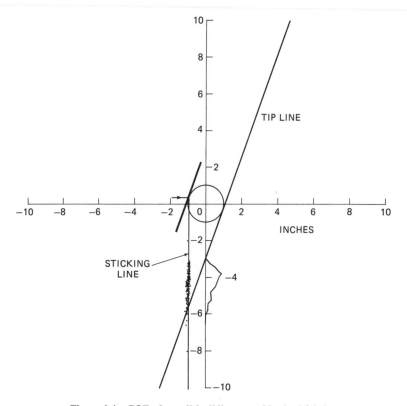

Figure 6-4　CORs for a disk sliding on rubberized fabric.

The fence angle at which slipping begins may be identified as roughly 14.5 ± 2 degrees, corresponding to $\mu_c = .25 \pm .05$.

6.5. INTERACTION OF A WORKPIECE WITH THE ENDPOINT OF A FENCE

This experiment measured the final orientation of a brass square after it had come into alignment with a moving fence, slid to the end of the fence, and turned off the end. As illustrated in Figure 6-9, the calculated final orientations may vary from somewhat more than the fence angle up to 90 degrees.

　　At each fence angle from 25 to 65 degrees (at 5-degree intervals), the robot performed 50 pushing operations. Sliding surfaces were rubberized fabric (Figure 6-10), 180 grade emery cloth (Figure 6-11), and steel (Figure 6-12). Fence speed was 2 cm/sec. In the figures, the calculated bounds on the final angle are shown as solid curves. These bounds were computed for $\mu_c =$

.25 as measured. A histogram of the number of times a given final angle was observed is plotted horizontally. In all cases the final angles fall well within the calculated bounds.

Some bimodal behavior can be observed in many of the histograms. This probably results from the fact that the square was rotated by 90 degrees after each operation. An oscillation of period 2 can be observed in the final orientations when listed sequentially. (The sequence information is lost in the histograms.) Presumably there is some difference in the surface characteristics of the brass square's orthogonal sides.

One might ask why the final orientations of this endpoint operation obey calculated bounds even on sliding surfaces (such as steel) which gave

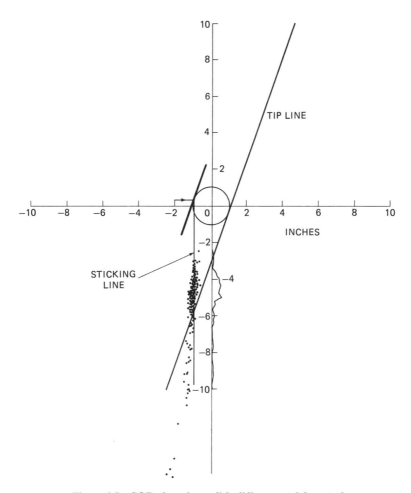

Figure 6-5   CORs for a brass disk sliding on stainless steel.

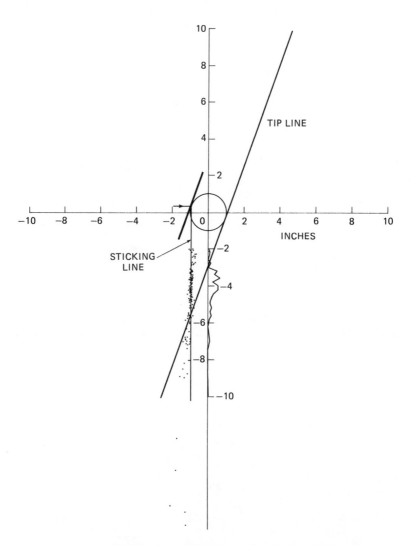

Figure 6-6   CORs for a brass disk sliding on lubricated stainless steel.

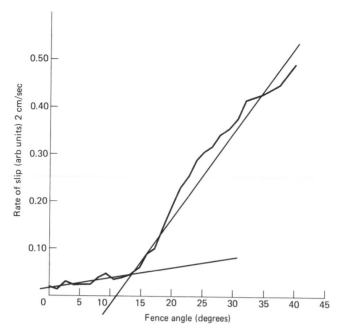

Figure 6-7   Amount of slipping along the fence for various fence angles, at 2 cm/ sec. To find the coefficient of contact friction μ_c between the aluminum fence and the edge of a brass workpiece, the amount of slip is plotted here against the fence angle $(90 - \alpha)$. The best estimate of μ_c is the tangent of the angle at the intersection of the two linear fits.

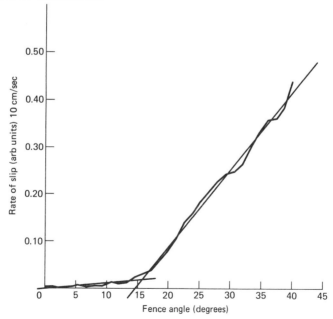

Figure 6-8   Amount of slipping along the fence for various fence angles, at 10 cm/ sec.

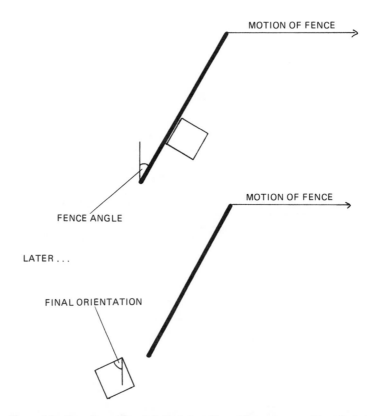

MOTION OF FENCE

FENCE ANGLE

MOTION OF FENCE

LATER . . .

FINAL ORIENTATION

Figure 6-9 Experiment for studying interaction of brass square with endpoint of fence. To test the predictions of gross motion of a workpiece as it interacts with a fence, this experiment was performed repeatedly. The fence translates horizontally, and its front edge is angled at the angle $90 - \alpha$ shown. The final orientation of a square workpiece which has slid along the fence and off the end is measured and compared with calculations.

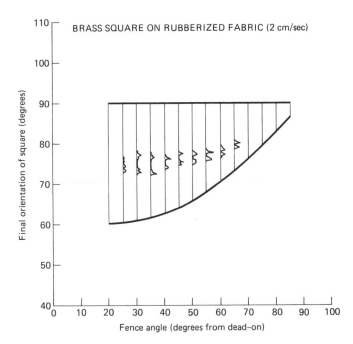

Figure 6-10   Final orientations of the square versus fence angle, on rubberized fabric. At the fence angles $90 - \alpha$ shown (5-degree increments), the final orientation of the pushed square is plotted as a histogram horizontally. Calculated bounds are the upper straight line and the lower curve. All the final orientations at all fence angles lie well within the calculated bounds.

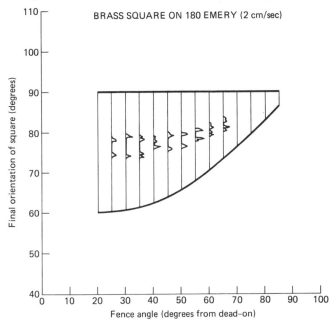

Figure 6-11   Final orientations of the square versus fence angle, on emery cloth.

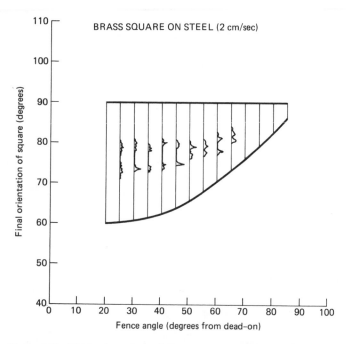

Figure 6-12   Final orientations of the square versus fence angle, on steel.

poor agreement in the sticking locus experiment (Section 6.3). Probably this is because the much greater sliding distance in this experiment gave any momentary violation of the bounds a chance to be averaged out.

6.6. CW VERSUS CCW ROTATION

The parts-feeders described in Section 5.4 depend on final angles after an endpoint interaction to fall within calculated bounds as tested, and also require workpieces to turn clockwise or counterclockwise as determined by Mason's rules [44] (Section 1.6.2).

In this experiment the corner of a brass square is struck by a fence at an angle near the critical angle dividing CW rotation from CCW rotation. Figure 6-13 shows the geometry of the experiment. The initial angle β_o of the CM of the square with respect to the line of motion is varied in the vicinity of 0 degrees. After pushing about 5 cm, the final angle β_f of the square is measured to see if it has turned CW or CCW. Results are plotted in Figure 6-14. At each initial angle of motion β_o (at 1-degree intervals) five trials were performed. The sliding surface was rubberized fabric. As can be seen from the figure, correct rotation can be assured only if the initial angle

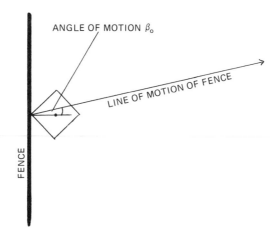

Figure 6-13 Experiment for studying CW versus CCW rotation. In this experiment a fence with vertical front edge ($\alpha = 90$) translates a few degrees from horizontal. This is equivalent to varying α a few degrees from 90. The pushed workpiece was initially oriented so that if the fence translated exactly horizontally, the workpiece should rotate neither CW nor CCW. The amount the workpiece actually did rotate after being pushed about 5 cm is measured and compared to the angle of motion of the fence.

of motion is at least 2 degrees from the critical angle (in this case $\beta_o = 0$) dividing CW from CCW.

In addition to the 2-degree uncertainty, there is a systematic shift of about 1 degree in initial angle of motion β_o. At most only about half of this can be attributed to systematic misalignment of the square. A more extreme systematic shift is seen in Figure 6-15, in which one trial was performed at each angle on emery cloth. The emery cloth had been used in previous experiments, and brass dust was heavy on some areas of it, changing its coefficient of friction. By chance the present experiment was performed on a part of the emery cloth in which the brass dust was highly nonuniform. A 5-degree systematic shift in initial angle of motion resulted. Apparently nonuniform μ_s can be a significant problem.

6.7. COR LOCUS

In Section 6.3, where the sticking locus was found, we arranged for μ_c to be sufficiently high that sticking always occurred. Thus the two-dimensional COR locus was compressed to a one-dimensional sticking locus. The other extreme is to make μ_c zero, so that the only force the fence can exert on the workpiece is one normal to the fence surface. In this case a two-dimensional COR locus results.

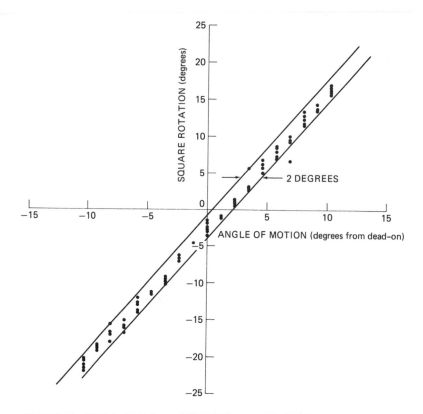

Figure 6-14   Final angle β_f versus initial angle β_o on rubberized fabric. According to Mason's rules for direction of rotation, the workpiece should switch from CW rotation to CCW rotation at zero degrees on this graph, that is, the graph should be symmetric about the y axis. In fact, correct rotation direction is only observed reliably when the angle of motion differs by 2 degrees or more from the nominal direction separating CW from CCW rotation.

It is difficult to obtain a fence with $\mu_c = 0$. But the same effect can be obtained by pulling on the corner of the workpiece with a string, as shown in Figure 6-16. A string can exert no tangential forces, but only a force along its length. In Figure 6-17 330 CORs obtained in this way are plotted, along with the calculated bounds. The equivalent fence angle $\alpha = 77.4$ degrees. These CORs were obtained on a 180 grade emery cloth surface, with pulls of length approximately 1 cm and speed 2 cm/sec.

As in the sticking locus experiment on the same sliding surface, the observed CORs are considerably smaller in distribution than the bounds would allow.

6.8. CHARACTERISTIC SPEED FOR QUASI-STATIC MOTION

We have assumed that the relative velocities of workpieces and their pushers are sufficiently small that frictional forces dominate inertial forces. Here we find characteristic velocities for which this "quasi-static" assumption is valid. Another approach to this problem is given in [43]. The opposite extreme, where inertial forces dominate frictional ones, has been treated in [67] (see Section 7.4).

To identify a characteristic velocity, we must specify the permissible error which neglect of inertial forces causes. For a given permissible error, an upper bound on velocity can be found such that at all lower velocities the permissible error is not exceeded. Permissible angular error will be denoted

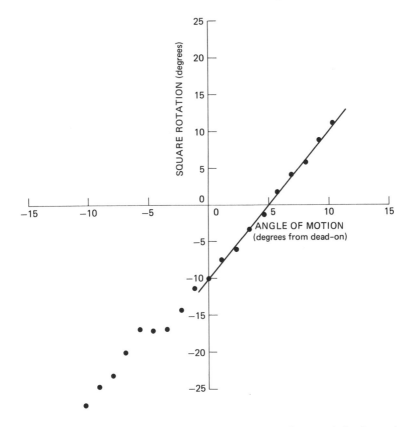

Figure 6-15 Final angle β_f versus initial angle β_o on used emery cloth. On used emery cloth, filled with brass filings from previous experiments, the coefficient of sliding friction μ_s is nonuniform, violating one of our assumptions. We see that the angle of motion dividing CW rotation from CCW rotation differs by about 5 degrees from its theoretical value of zero.

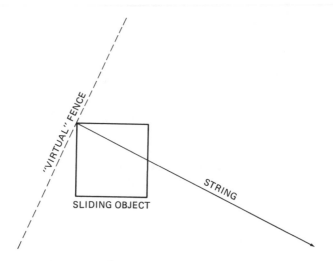

Figure 6-16 Experiment equivalent to pushing with a $\mu_c = 0$ fence. In this experiment a workpiece is pulled by a string at a known angle, which is equivalent to pushing it with a frictionless fence perpendicular to the fence. In either case the applied force direction is known (it lies along the string), and a two-dimensional distribution of CORs should result.

θ, while permissible error in displacement will be denoted X. The characteristic velocity depends on the details of the collision considered. Since we desire only an approximate upper bound for velocities, we will consider only a couple of typical systems.

6.8.1. Angular Error

Consider the "endpoint" problem (Section 5.1) in which a square turns as it leaves the endpoint of a fence with which it was aligned. Bounds were found on its rotation in the quasi-static limit. At nonzero velocities, how much can it rotate due to its angular momentum?

If the relative linear velocity of the fence and workpiece is v, the angular velocity of the workpiece after interaction is $\omega = v/a$, where a is the radius of the workpiece (or of the disk which circumscribes it). The rotational energy in the workpiece is then

$$E_{rot} = \frac{1}{2} I \omega^2 \tag{6.1}$$

where

$$I = M \frac{a^2}{2}$$

is the moment of inertia of the workpiece, in this case taken to be a disk of uniform density.

The disk will rotate an angle θ until the rotational energy is spent. The energy lost to friction during this rotation is found to be

$$E_{fric} = \frac{2}{3} \mu_s \theta a M g \tag{6.2}$$

assuming a uniform pressure distribution. g is gravitational acceleration. Equating the energies we find

$$v^2 = \frac{8}{3} \mu_s a g \theta \tag{6.3}$$

as a characteristic velocity. In the case of a workpiece with $a = 3.5$ cm and $\mu_s = .25$, errors of $\theta = 5$ degrees occur when $v = 14$ cm/sec.

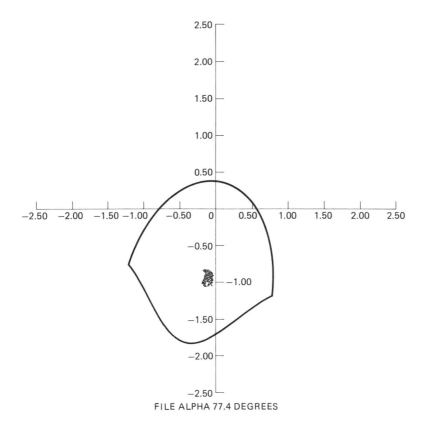

FILE ALPHA 77.4 DEGREES

Figure 6-17 CORs observed for $\mu_c = 0$ on emery cloth. The CORs observed are plotted as dots. They all fall well within the calculated boundary.

6.8.2. Translational Error

Here we find the kinetic energy and energy lost to friction in distance X to be

$$E_{trans} = \frac{1}{2} M v^2 \tag{6.4}$$

$$E_{fric} = M g \mu_s X \tag{6.5}$$

respectively. Equating the energies we find

$$v^2 = 2 g \mu_s X \tag{6.6}$$

as a characteristic velocity. If permissible error $X = .5$ cm, we find $v = 16$ cm/sec.

6.8.3. A Quasi-static Parameter

If we set the rotational error θ or the translational error X to values typical of the whole rotation or translation occurring in an interaction, we can find a characteristic velocity at which kinetic effects and quasi-static effects have similar magnitude. We use $\theta = 1$ (radian) and $X = a$. From either equation (6.3) or (6.6) we obtain roughly

$$v^2 = 2 \mu_s a g \tag{6.7}$$

as a characteristic velocity. With $\mu_s = .25$ and $a = 3.5$ cm, we have $v = 40$ cm/sec.

7

Suggestions for Further Work

7.1. USE IN CONJUNCTION WITH SENSORS

We have not treated the possibility of integrating the planning methods described here with information from sensors. In Section 1.1.3 we mentioned three distinct forms such integration could take:

1. Sensorless manipulation can be used below the resolution limit of sensors.
2. The mechanics of sliding and appropriate planning methods can be used to optimize progress toward a goal state from a sensed present state, thus making less frequent sensory measurements necessary.
3. Sensory information can be used to narrow the range of initial states a sensorless strategy must contend with.

As an example of the third form, in designing a conveyor belt–based parts-feeder, we might be unable to find a sequence of fences which will align a particular workpiece. A sensor with a simple yes/no output might detect a feature of the workpieces, and control a movable fence, making a parts-feeder possible. (In fact, a system having such a sensor and movable fence has been constructed by Pherson, Boothroyd, and Dewhurst [56], though no prediction of sliding motion was used.) In terms of the configuration maps used to plan the feeder, the presence of the sensor effectively removes some possible initial orientations. Consequently the range of possible orientations at each node of the search tree is reduced, and some of the nodes may become goal states (having only one possible orientation or a narrow range).

Simple sensory information can also be integrated into an otherwise sensorless robot strategy. The integrated strategy may replace use of more complex sensors. For instance, a grasping strategy for a robot may involve alignment of a face of a workpiece with a gripper surface. To avoid causing misalignment of the workpiece tangential to the gripper face, we might wish the workpiece not to slip relative to the gripper while it is being aligned. A slip sensor on the gripper face could control robot motion to prevent slip.

7.2. REALISTIC MODELS OF FRICTION

An important assumption used in deriving the COR loci is that of a coefficient of sliding friction μ_s which is uniform over the sliding surface and velocity independent: simple Coulomb friction. Friction is rarely so well behaved.

Velocity dependence of μ_s will have only moderate consequences for the COR loci. The sense of rotation (CW or CCW) is not affected by velocity dependence, because pure translation of the workpiece is the marginal case dividing the senses of rotation. In pure translation all parts of the workpiece move with the same velocity, so velocity dependence of μ_s is unseen by the workpiece. If μ_s decreases with increasing velocity (the usual case), we can predict that CORs will lie closer to center of mass (i.e., rotation rates will be faster) than they would with constant μ_s. The side of the workpiece toward which it turns has a lower velocity, therefore higher μ_s, and therefore more drag, causing the workpiece to turn still faster toward that side.

Spatial nonuniformity of μ_s is more serious. In Section 1.6.2 a nonuniformly worn surface caused a 5-degree offset in the marginal pushing direction dividing CW from CCW rotation. It would be hard to control such a major effect analytically. Instead, sliding surfaces must be kept uniform.

When it is the surface of the sliding workpiece, rather than the surface of the table, which is nonuniform, we may hope to find simple analytic adjustments to the COR locus to compensate. The distinction between *center of friction* (*COF*) and *center of mass* becomes important [39] [44]. If the composition of the workpiece surface is understood and some information about the pressure distribution is available, a COF distinct from the CM can be calculated for the workpiece. Then the sense of rotation (at least) will be predictable. It is not known what effect a COF distinct from the CM will have on the COR locus.

7.3. PUSHING ABOVE THE PLANE

We assume that the point of contact between pusher and pushed workpiece is not far (relative to the radius of the workpiece) above the sliding surface. In the extreme case of pushing far above the plane, the workpiece will tip over instead of sliding. For small heights, the effect creates a center of friction (COF) distinct from the center of mass (CM). The effect on the COR locus is unknown.

7.4. NON-QUASI-STATIC VELOCITIES

The results of Chapters 3 and 4 depend on the quasi-static assumption, discussed in Section 2.1.1. We assume that dissipative effects due to friction between a workpiece and the surface it slides on overwhelm inertial effects. In the real world both effects are present, and become of comparable importance at characteristic speeds calculated in Section 6.8 and in [43]. The results of Chapters 3 and 4 may be considered to be the $\nu \to 0$ limit.

In the opposite extreme we may neglect sliding friction altogether, and only consider inertial effects. The motion is then independent of speed, so we may consider this case to be the $\nu \to \infty$ limit. The details of the contact between the sliding workpiece and the surface it slides on (the pressure distribution, Section 3.1.5) no longer have any effect on the motion. For given initial conditions then, a *single* resulting motion can be calculated rather than the *locus* of possible motion calculated for slow motions.

Drawing on the work of Routh [61], Wang [67] has calculated the motion of a pushed workpiece in the $\nu \to \infty$ limit (the "impact" limit). The motion is a function of the coefficient of friction between the pusher and workpiece μ_c as in the quasi-static case and of the geometry of pushing, but it also depends on the elasticity of the materials in contact. Elasticity ranges from the plastic limit $e = 0$ (e.g., modeling clay) to the elastic limit $e = 1$ (e.g., spring steel).

The instantaneous motion on impact can be described by a center of rotation (COR) somewhere in the plane. Wang finds [68] that when $e = 0$ or $\mu_c = 0$, the COR falls along the axis of symmetry of the quasi-static COR locus derived in Chapter 3 and at a distance r_{impact} from the CM given by

$$r_{impact} = \frac{\rho^2}{\vec{\alpha} \cdot \vec{c}} \tag{7.1}$$

where ρ is the radius of gyration of the workpiece. Equation (7.1) is the same as equation (3.37) (which gives the tip of the COR locus in the quasi-static case) when the workpiece pushed is a circular rim, for which $\rho = a$.

For all other workpieces $\rho < a$, so we may conclude that the COR for impact lies within the COR locus for quasi-static pushing, if $e = 0$ or $\mu_c = 0$.

If $e > 0$ and $\mu_c > 0$, Wang finds that the sense of rotation (CW or CCW) does not necessarily agree with Mason's results for quasi-static motion (summarized in Section 1.6.2). This means that in realistic cases where $e > 0$, a given sliding operation which results in CW rotation of the pushed workpiece in the quasi-static limit may change over to CCW rotation as velocity is increased.

For fixed elasticity e and coefficients of friction μ_c and μ_s, as velocity is increased the locus of CORs describing the motion must change continuously from the quasi-static locus at $v = 0$ to the single point (sometimes outside the quasi-static locus) which is Wang's result at $v = \infty$. If the COR loci for intermediate velocities could be found or bounded, motion-planning algorithms based on sliding friction (e.g., Chapter 5 and references [12] and [40]) could be extended to non-quasi-static velocities.

7.5. BOUNDS ON THE COR LOCUS

The COR loci found in Chapter 3 are exact if the sliding workpiece is a disk. Any COR in the locus could occur for some combination of bumps on the bottom of a disk, that is, for some pressure distribution. The COR locus for a disk necessarily encloses the COR locus for any workpiece which could be enclosed in that disk, with the same center of mass. The COR locus for the inscribed workpiece may be considerably smaller than that for the disk, especially when the area of the inscribed workpiece is considerably less than that of the disk. The COR locus for a square, found numerically, and the outline of the COR locus for a disk circumscribing the square are shown in Figure 3-6.

For comparison, the line of CORs for a *uniform* pressure distribution on a disk is shown in Figure 7-1. In the uniform case, for each α (related to the force angle) there is of course only one COR, as the pressure distribution is completely specified. The COR for uniform pressure distribution was found by numerical integration.

Shown in the figure is a particular α_1 for illustration, and the COR locus outline for all pressure distributions for that α_1. The tip line for all α is shown. The point of intersection of the $\vec{\alpha}_1$ vector through the CM and the tip line is indicated with a dot, which is the tip of the COR locus for α_1. The tip of the COR locus for *any* α lies at the intersection of the $\vec{\alpha}$ vector through the CM and the tip line.

Similarly, the COR for uniform pressure for α_1 lies at the intersection of the $\vec{\alpha}_1$ vector through the CM and the uniform pressure line, as indicated by a dot. The COR for uniform pressure for *any* α lies at the intersection of the $\vec{\alpha}$ vector through the CM and the line of uniform pressure.

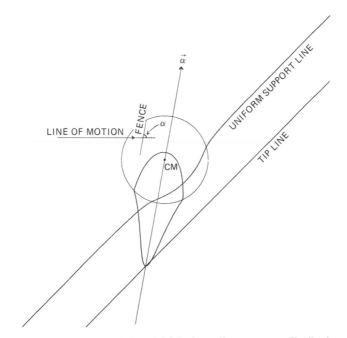

Figure 7-1   **Tip line and line of CORs for uniform pressure distribution.**

Using the COR loci of disks in planning manipulation strategies for other shapes results in unnecessarily conservative strategies. It is even possible that no strategy might be found when one exists. This problem could be alleviated if exact COR loci for arbitrarily shaped workpieces could be found. In finding the COR locus for a disk, we discovered two classes of "dipods" (pressure distributions consisting of only two points of support, Section 3.3.5) which were responsible for the boundary of the COR locus. For workpieces other than disks, the boundary is not described by dipods, and finding the COR locus becomes considerably harder.

7.6. USING CONFIGURATION MAPS

In Chapter 5 *configuration maps* are introduced, describing elementary operations on a workpiece to be manipulated. The elementary operations for which configuration maps are actually computed are interactions of the workpiece with a fence. These maps turn out to have a simple form, being relations of bounded sets of orientations. We can take advantage of that form by rectangularizing the map and encoding it symbolically (Section 5.2.1). The computation of a product of several maps (which is the map for a sequence of elementary operations) then becomes trivial, and searching the space of operation sequences is greatly facilitated.

For general operations configuration maps will not be made up of rectangles, so rectangularizing the configuration map will result in overly conservative planning. The harm may be minimized by a good covering of the configuration maps by simple shapes. This is a problem closely related to the representation of unoccupied areas of C-space [11] [9] [10].

7.7. POSSIBILITY AND PROBABILITY IN CONFIGURATION MAPS

COR loci and configuration maps are based on possibility, not probability. The COR locus gives the set of all possible instantaneous motions of a pushed workpiece, but no indication of the relative probability of each. To calculate the relative probability of CORs within the COR locus would require a model of the pressure distribution, and therefore of the surfaces in

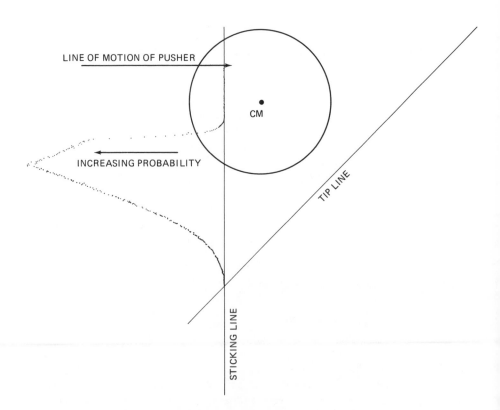

Figure 7-2   Histogram of CORs along sticking line, for random tripods of support.

contact, which is exactly what we set out to avoid since that information is generally unavailable.

Even if we model the pressure distributions as randomly selected tripods (pressure distributions having but three nonzero points, discussed in Section 3.3.3), finding the COR for a given tripod is surprisingly difficult. Some of the causes for the difficulty are discussed in [44]. To generate the locus shown in Figure 3-5, we were compelled to generate each tripod from a previous one by moving one "foot" a minute distance and starting the minimization from the COR found for the previous tripod. Otherwise the minimization technique used did not converge, and even so it did not always converge. (A slower but more reliable method was used in [44].) While adequate for finding the COR locus, the biased set of tripods so selected does not randomly sample the set of all tripods.

However, if the coefficient of friction between workpiece and pusher μ_c is taken to be large, the COR locus collapses onto the sticking line (Chapter 4). Now the search for a COR for a given tripod can be restricted to a one-dimensional range rather than the two-dimensional range for $\mu_c = 0$ in which minimization was difficult. A histogram of the relative probabilities of various CORs along the sticking line, for randomly selected tripods, is shown in Figure 7-2.

Configuration maps (Section 5.2) were taken to be mappings from two copies of C-space (C-space × C-space) into 0 (not possible) or 1 (possible). They could map into the interval [0, 1] instead, representing probability, if, as just discussed, probability information is available.

References

1. Aristotle. *Physics*.

2. H. Asada. *Studies on Prehension and Handling by Robot Hands with Elastic Fingers*. Ph.D. Th., Kyoto University, April 1979.

3. H. Asada and A. B. By. Kinematics of Workpart Fixturing. Proceedings, IEEE Int'l Conf. on Robotics and Automation, St. Louis, MO, March 1985, pp. 337–345.

4. B. S. Baker, S. Fortune, and E. Grosse. Stable Prehension with a Multi-fingered Hand. Proceedings, IEEE Int'l Conf. on Robotics and Automation, St. Louis, MO, March 1985.

5. J. Barber, R. A. Volz, R. Desai, R. Rubinfield, B. Schipper, and J. Wolter. Automatic Two-fingered Grip Selection. Proceedings, IEEE Int'l Conf. on Robotics and Automation, San Francisco, CA, April 1986, pp. 890–896.

6. H. C. Berg. "How Bacteria Swim." *Scientific American 233*, 2 (August 1975).

7. G. Boothroyd, C. Poli, and L. E. Murch. *Automatic Assembly*. New York: Marcel Dekker, 1982.

8. Rodney A. Brooks. "Symbolic Error Analysis and Robot Planning." *International Journal of Robotics Research 1*, 4 (Winter 1982), pp. 29–68.

9. R. A. Brooks. "Solving the Find-Path Problem by Good Representation of Free Space." *IEEE Transactions on Systems, Man, and Cybernetics SMC-13*, 13 (1983), pp. 190–197.

10. R. A. Brooks and T. Lozano-Perez. A Subdivision Algorithm in Configuration Space for Findpath with Rotation. Proceedings, 8th International Joint Conference on Artificial Intelligence, August 1983, pp. 799–806.

11. R. A. Brooks. "Planning Collision-Free Motions for Pick-and-Place Operations." *The International Journal of Robotics Research 2*, 4 (Winter 1983), pp. 19–44.

12. Randy Brost. Automatic Grasp Planning in the Presence of Uncertainty. Proceedings, IEEE Int'l Conf. on Robotics and Automation, San Francisco, CA, April 1986, pp. 1575–1581.

13. M. Cutkosky. Personal communication.

14. Mark R. Cutkosky. Mechanical Properties for the Grasp of a Robotic Hand. TR 84-24, Robotics Institute, Carnegie-Mellon University, Pittsburgh, PA, Robotics Inst., September 1984.

15. Mark R. Cutkosky. *Grasping and Fine Manipulation for Automated Manufacturing.* Ph.D. Th., Carnegie-Mellon University, January 1985.

16. M. R. Cutkosky and P. K. Wright. "Active Control of a Complaint Wrist in Manufacturing Tasks." *ASME Journal of Engineering for Industry 108,* 1 (1986), pp. 36–43.

17. Michael Andreas Erdmann. On Motion Planning with Uncertainty. Master Th., AI-TR-810. Massachusetts Institute of Technology, Cambridge, 1984.

18. M. A. Erdmann and M. T. Mason. An Exploration of Sensorless Manipulation. Proceedings, IEEE Int'l Conf. on Robotics and Automation, San Francisco, CA, April 1986, pp. 1569–1575.

19. R. S. Fearing. Simplified Grasping and Manipulation with Dextrous Robot Hands. American Control Conference, San Diego, CA, June 1984, pp. 32–38.

20. Herbert Goldstein. *Classical Mechanics,* 2nd ed. Reading, MA: Addison-Wesley, 1980.

21. B. R. Gossick, "A Lagrangian Formulation for Nonconservative Linear Systems Which Satisfies Hamilton's Principle." *IEEE Transactions on Education E-10,* 1 (1967), pp. 37–42.

22. Suresh Goyal, Andy Ruina, and Jim Papadopoulos. "Planar Sliding with Dry Friction 1: Limit Surface and Moment Function" and "Planar Sliding with Dry Friction 2: Dynamics of Motion." Submitted to WEAR.

23. D. D. Grossman and M. W. Blasgen. "Orienting Mechanical Parts by Computer Controlled Manipulator." *IEEE Transactions on Systems, Man, and Cybernetics SMC-5,* 5 (September 1975), pp. 561–565.

24. W. Holzmann and J. M. McCarthy. On Grasping Planar Objects with Two Articulated Fingers. Proceedings, IEEE Int'l Conf. on Robotics and Automation, St. Louis, MO, March 1985, pp. 576–581.

25. H. Inoue. Force Feedback in Precise Assembly Tasks. MIT AI Memo 308, Massachusetts Institute of Technology, Cambridge, August 1974.

26. John W. Jameson. *Analytic Techniques for Automated Grasp.* Ph.D. Th., Stanford University, Stanford, CA, June 1985.

27. J. W. Jameson and L. J. Leifer. Quasi-Static Analysis: A Method for Predicting Grasp Stability. Proceedings, IEEE Int'l Conf. on Robotics and Automation, San Francisco, CA, April 1986, pp. 876.

28. J. H. Jellett. *A Treatise on the Theory of Friction.* London: Macmillan, 1872.

29. J. Kerr. *Issues Related to Grasping by Robots.* Ph.D. Th., Stanford University, Stanford, CA, January 1985.

30. J. Kerr and B. Roth. "Analysis of Multifingered Hands." *The International Journal of Robotics Research 4,* 4 (1986).

31. L. D. Landau and E. M. Lifshitz. *Mechanics,* 3rd ed. Oxford: Pergamon Press, 1976.

32. Christian Laugier. A Program for Automatic Grasping of Objects with a Robot Arm. Proceedings, 11th Int'l Symposium of Industrial Robots, Soc. Biomechanisms Japan & Japan Industrial Robot Assoc., Tokyo, 1981, pp. 287–294.

33. Zexiang Li and Shankar Sastry. Task-Oriented Optimal Grasping by Multi-fingered Robot Hands. TR UCB/ERL M86/43, Electronics Research Lab, University of California, Berkeley, May 1986.

34. J. Loncaric. *Normal Forms of Stiffness and Compliance Matrices.* Ph.D. Th., Harvard University, Cambridge, MA, March 1986.

35. Tomas Lozano-Perez. The Design of a Mechanical Assembly System. Master Th., Massachusetts Institute of Technology, Cambridge, December 1976.

36. Tomas Lozano-Perez. "Spatial Planning: A Configuration Space Approach." *IEEE Transactions on Computers C-32,* 2 (February 1983), pp. 108–120.

37. T. Lozano-Perez, M. T. Mason, and R. H. Taylor. "Automatic Synthesis of Fine-Motion Strategies for Robots." *International Journal of Robotics Research 3,* 1 (Spring 1984), pp. 3–24.

38. Tomas Lozano-Perez. "Motion Planning and the Design of Orienting Devices for Vibratory Parts Feeders." *IEEE Journal of Robotics and Automation* (1986).

39. W. D. MacMillan. *Dynamics of Rigid Bodies.* New York: Dover, 1936.

40. M. Mani and W. R. D. Wilson. A Programmable Orienting System for Flat Parts. Proceedings, NAMRII XIII, 1985.

41. Matthew T. Mason. "Compliance and Force Control for Computer Controlled Manipulators." *IEEE Transactions on Systems, Man, and Cybernetics SMC-11,* 6 (June 1981), pp. 418–432.

42. Matthew T. Mason. *Manipulator Grasping and Pushing Operations.* Ph.D. Th., AI-TR-690. Massachusetts Institute of Technology, Cambridge, June 1982.

43. Matthew T. Mason. On the Scope of Quasi-static Pushing. Proceedings, 3rd Int'l Symposium on Robotics Research, October 1985.

44. Matthew T. Mason. "Mechanics and Planning of Manipulator Pushing Operations." *International Journal of Robotics Research 5,* 3 (Spring 1986).

45. M. T. Mason and R. Brost. Automatic Grasp Planning: An Operation Space Approach. Proceedings, 6th Symposium on the Theory and Practice of Robots and Manipulators, Cracow, Poland, September 1986.

46. B. K. Natarajan. *On Moving and Orienting Objects.* Ph.D. Th., Computer Science Dept., Cornell University, Ithaca, NY, August 1986.

47. Van-Duc Nguyen. The Synthesis of Stable Grasps in the Plane. Proceedings, IEEE Int'l Conf. on Robotics and Automation, San Francisco, CA, April 1986, p. 884.

48. N. Nilsson. A Mobile Automaton: An Application of Artificial Intelligence Techniques. 2nd International Joint Conference on Artificial Intelligence, 1969.

49. H. Ozaki, A. Mohri, and M. Takata. "On the Force Feedback Control of a Manipulator with a Compliant Wrist Force Sensor." *Mechanism and Machine Theory (GB) 18,* 1 (1983), pp. 57–62.

50. M. A. Peshkin and A. C. Sanderson. The Motion of a Pushed, Sliding Object. Part 1: Sliding Friction, Tech. Rept. CMU-RI-TR-85-18. Robotics Institute, Carnegie-Mellon University, Pittsburgh, PA, September, 1985.

51. M. A. Peshkin and A. C. Sanderson. The Motion of a Pushed, Sliding Object. Part 2: Contact Friction, Tech. Rept. CMU-RI-TR-86-7. Robotics Institute, Carnegie-Mellon University, Pittsburgh, PA, April 1986.

52. M. A. Peshkin and A. C. Sanderson. Manipulation of a Sliding Object. Proceedings, IEEE Int'l Conf. on Robotics and Automation, San Francisco, CA, April 1986, pp. 233–239.

53. M. A. Peshkin and A. C. Sanderson. Planning Robotic Manipulation Strategies for Sliding Objects. Proceedings, IEEE Int'l Conf. on Robotics and Automation, Raleigh, NC, April 1987, pp. 696–701.

54. Michael A. Peshkin. *Planning Robotic Manipulation Strategies for Sliding Objects*. Ph.D. Th., Physics Department, Carnegie-Mellon University, Pittsburgh, PA, October 1986.

55. M. A. Peshkin and A. C. Sanderson. "Reachable Grasps on a Polygon: The Convex Rope Algorithm." *IEEE Journal of Robotics and Automation RA-2, 1* (March 1986), pp. 53–58.

56. D. Pherson, G. Boothroyd, and P. Dewhurst. Programmable Feeder for Nonrotational Parts. In W. B. Heginbotham, ed., *Programmable Assembly*. IFS (Publications), London, 1984, pp. 247–256.

57. K. Pingle, R. Paul, and R. Bolles. *Programmable Assembly: Three Short Examples*. Film, Stanford AI Lab, Stanford, CA, 1974.

58. J. Prescott. *Mechanics of Particles and Rigid Bodies*. London: Longmans, Green, 1923.

59. M. H. Raibert and J. J. Craig. "Hybrid Position/Force Control of Manipulators." *ASME Journal of Dynamic Systems, Measurement, and Control 102* (June 1981).

60. M. Brady et al., eds. *Robot Motion, Planning and Control*. Cambridge, MA: The MIT Press, 1982.

61. E. J. Routh. *Dynamics of a System of Rigid Bodies,* 7th ed. New York: Dover, 1960.

62. A. Ruina. Personal communication.

63. M. T. Mason and J. K. Salisbury. *Robot Hands and the Mechanics of Manipulation*. Cambridge, MA: The MIT Press, 1985.

64. S. N. Simunovic. Force Information in Assembly Processes. Proceedings, 5th International Symposium on Industrial Robots, September 1975.

65. J. Trinkle. *The Mechanics and Planning of Enveloping Grasps*. Ph.D. Th., University of Pennsylvania, Philadelphia, 1987.

66. S. Udupa. Collision Detection and Avoidance in Computer Controlled Manipulators. 5th International Joint Conference on Artificial Intelligence, Cambridge, 1977.

67. Yu Wang. On Impact Dynamics of Robotic Operations. Tech. Rept. CMU-RI-TR-86-14, Robotics Institute, Carnegie-Mellon University, Pittsburgh, PA, 1986.

68. Yu Wang. Personal communication.

69. D. E. Whitney. "Force Feedback Control of Manipulator Fine Motions." *ASME Journal of Dynamic Systems, Measurement, and Control* (June 1977), pp. 91–97.

70. D. E. Whitney. "Discrete Parts Assembly Automation—An Overview." *ASME Journal of Dynamic Systems, Measurement, and Control* (March 1979).

71. D. E. Whitney. "Quasi-Static Assembly of Compliantly Supported Rigid Parts." *ASME Journal of Dynamic Systems, Measurement, and Control 104* (March 1982), pp. 65–77.

72. D. E. Whitney. "Quasi-Static Assembly of Compliantly Supported Parts." *ASME Transactions on Dynamic Systems, Measurement and Control 104* (March 1982), pp. 65–77.

73. Jan D. Wolter, Richard A. Volz, and Tony C. Woo. "Automatic Generation of Gripping Positions." *IEEE Transactions SMC, 15,* 2, (March/April 1985), pp. 204–213.

Index

Robot (*cont.*)
  rigidity of, quasi-static approximation and, 27

Search tree, 19, 109–12
  pruning of, 19, 112
Sensing, and uncertainty, 3–4
Single three-point pressure distribution, 12
Sliding:
  center of rotation (COR), 11–12
  Mason's rules for rotation, 10–11
  physical effect, 9–10
  physics of, 20–21
  planar sliding, 22
Sliding friction, 5, 13, 44
Sliding workpiece, experiments, 115
Sliding workpiece (COR locus), 14–16, 41–72
  analytic form of COR locus, 58–61
  analytic solution, 50–52
  application, 69–72
  $|COR| < a$, 55–56, 61–64
  $|COR| > a$, 56–58
  energy lost to friction with the table, 48–49
  iterative numerical solution, 49–50
  minimum power principle, 46–47
  motion of pusher/rotation of workpiece, relation between, 47–48
  quotient locus:
    extrema of, 52–54
    numerical exploration of, 54–55
  range of applicability, 41–46
    bounding the workpiece by a disk, 45
    center of rotation (COR), 42–43
    Coulomb friction, 44
    geometric parameters, 45–46
    notation, 46
    point of contact between workpiece and pusher, 42
    position controlled pusher, 42
    pressure distribution between workpiece and table, 43–44
    quasi-static motion, 44
    workpiece shape, 41
  solution, 47–69
  symmetries of COR locus, 68
Slipping, consistency for, 77–78
Slipping-slowest behavior, 85–86, 87

transition from sticking-slowest behavior to, 86
Slipping zones, 76–77
Spiral localization (disk), 96–104
  analysis, 96–97
  critical radius versus collision parameter, 100
  fastest guaranteed spiral, computation of, 102–4
  limiting radius, 100–102
  pusher chasing disk around a circular path, 97–100
Squeeze grasp, 9, 12
Sticking line, 75, 76, 79
  construction of, 76
Sticking locus, 77–79
  experiments, 116–18
Sticking-slowest behavior, 85, 87–88
  transition to slipping-slowest behavior, 86
Sufficiently sensor-intensive environments, 3–4
Symbolic encoding, configuration maps, 109

Tip line, 16, 65–68, 104
Tripods, 12, 46, 49

Uncertainty:
  compliant robot mechanics, 4–56
  and geometry, 3
  and physics, 2–3
  planning with, 1–4
  sensing, 3–4
Uncertainty reduction:
  in grasping, 21–22
  parts-feeders and, 8
Uniform pressure distribution, on disk, 136
Unit vector, 52
  notation, 46
Up-slipping locus, 77
Up-slipping zone, 75, 79
  construction of, 76

Velocity-dependent forces, 24–25
  minimum power principle and, 31
Virtual work, principle of, 26

Workpiece alignment, 8
Wrapped elementary configuration, 79–80